SpringerBriefs in Applied Sciences and Technology

SpringerBriefs present concise summaries of cutting-edge research and practical applications across a wide spectrum of fields. Featuring compact volumes of 50–125 pages, the series covers a range of content from professional to academic.

Typical publications can be:

- A timely report of state-of-the art methods
- An introduction to or a manual for the application of mathematical or computer techniques
- A bridge between new research results, as published in journal articles
- A snapshot of a hot or emerging topic
- An in-depth case study
- A presentation of core concepts that students must understand in order to make independent contributions

SpringerBriefs are characterized by fast, global electronic dissemination, standard publishing contracts, standardized manuscript preparation and formatting guidelines, and expedited production schedules.

On the one hand, **SpringerBriefs in Applied Sciences and Technology** are devoted to the publication of fundamentals and applications within the different classical engineering disciplines as well as in interdisciplinary fields that recently emerged between these areas. On the other hand, as the boundary separating fundamental research and applied technology is more and more dissolving, this series is particularly open to trans-disciplinary topics between fundamental science and engineering.

Indexed by EI-Compendex, SCOPUS and Springerlink.

More information about this series at http://www.springer.com/series/8884

Jeffrey Kok Hui Chan · Ye Zhang

Sharing by Design

Springer

Jeffrey Kok Hui Chan
Humanities, Arts and Social Sciences
Singapore University of Technology
and Design
Singapore, Singapore

Ye Zhang
Department of Architecture
National University of Singapore
Singapore, Singapore

ISSN 2191-530X ISSN 2191-5318 (electronic)
SpringerBriefs in Applied Sciences and Technology
ISBN 978-3-030-43568-4 ISBN 978-3-030-43569-1 (eBook)
https://doi.org/10.1007/978-3-030-43569-1

This Springer imprint is published by the registered company Springer Nature Switzerland AG
The registered company address is: Gewerbestrasse 11, 6330 Cham, Switzerland

For Heng Heng

Acknowledgements

The idea of this book first emerged from our many conversations as long-time colleagues and best friends on sharing, cities, ethics, and design at the Department of Architecture, National University of Singapore. These conversations evolved over a number of years, but they were all greatly enriched by the knowledge and insights of colleagues, friends, and family members along the way. Here, we would like to thank all of you.

The momentum of this book took shape during our participation in the "NUS-Tsinghua Design Research Initiative: Sharing Cities". And so, first and foremost, we would like to express our gratitude to Ng Teng Fong Charitable Foundation (Hong Kong). Without its sponsorship and support–both financial and intellectual–the "NUS-Tsinghua Design Research Initiative: Sharing Cities" would have been impossible, and likewise, also the several papers on space and sharing that we published, which subsequently paved the way to the birth of this book.

Our special gratitude also goes to our most generous, understanding, and supportive editors from Springer Nature: Megana Dinesh and Anthony Doyle. They received our book proposal with great enthusiasm, and our often abrupt requests were always received–and granted–with a generosity that greatly encouraged us to go further. Thank you, Megana and Anthony.

Here, we would like to extend our sincere thanks to Prof. Ho Puay Peng and Associate Professor Wong Yunn Chii from the Department of Architecture, National University of Singapore, and Prof. Li Xiaodong and Prof. Zhang Yue from the School of Architecture, Tsinghua University for their strong leadership in the "NUS-Tsinghua Design Research Initiative: Sharing Cities". We benefited immensely from their encouragement and support. We are especially grateful to our colleagues of this project, namely, Tan Teck Kiam from the National University of Singapore, Huang He, Martijn de Geus, Long Ying, and Liu Sha from Tsinghua University; and Seah Chee Huang, Tan Chee Yong, Brian Hwui Zhi Cheng, Shawn Eng Kiong Teo, Jonathan Christian Yi Ren Chin, and Karen Tan from DP Architects. Their collegiality, generosity, insights, and criticality helped to shape our thinking as we developed the ideas in this book.

Also, there are alway many supportive colleagues and friends behind even a small book such as this one. We are truly grateful and indebted to the following individuals during the preparation and writing of this book (listed in alphabetical order of the last name): Cheah Kok Ming; Lilian Chee; Cho Im Sik; Jason Woon Siong Chong; Simone Shu Yeng Chung; Heng Chye Kiang; Richard Kong Fatt Ho; Thomas Kong; Lam Khee Poh; Lim Sun Sun; Joseph Ee Man Lim; Lim Hwee Lee; Tomohisa Miyauchi; Jurgen Rosemann (& Vivienne Chiu Yuan Wang); Ruzica Stamenovic; Imran Bin Tajudee; Tan Beng Kiang; Teh Kem Jin (& Sylvia); Teo Yee Chin; and Zdravko Trivic.

We would like to thank all the students from the M.Arch program in the National University of Singapore and from the English Program of Master in Architecture, Tsinghua University, who participated in the NUS-Tsinghua Joint Design Studio on Sharing Cities from 2017 to 2020. Our thanks also go to those student who conducted their M.Arch theses under "NUS-Tsinghua Design Research Initiative: Sharing Cities", namely, Ong Guo Xiang and Gabriel Yong Woon Ng (2017); Kenny Han Teng Chen and Shi Yan Jie (2018); Yeow Yann Herng, Jessica Ho and Lam Ching Yan (2019). Our discussions and debates with all of you, as well as the hard questions that you posed, have sparked many important ideas and questions in this book. Moreover, He Zhuoshu, Research Assistant of the "NUS-Tsinghua Design Research Initiative: Sharing Cities" deserves special mention, especially for his invaluable work in searching and compiling the research materials during the preparation of this book.

Finally, we reserve our wholehearted thanks to each member of our family. From Jeffrey: Thank you, Felicity Hwee Hwa Chan (& Ji-Jon Sit), Dad and Mom. From Ye: my wife Zhuang Danshuo deserves the deepest and warmest thanks for her love, patience, and tolerance over the past year—not only for her forbearance with my lack of company during her pregnancy, but she also helped me going through the challenging period of writing with confidence. I would also like to sincerely thank my parents, whose support has always been the source of courage in my life.

Contents

List of Figures

List of Tables

Chapter 1
Introduction: Connecting Sharing and Design

Against a myriad of social and technological forces that drive people apart today, sharing is a relation that continues to bind people together. To share is to require the presence of another person. To share is also to create a new relation with this person—the creation of something defined as ours beyond what is yours and mine (Belk 2010). To share then is to cultivate a sense of togetherness (Jarvis 2019). These social possibilities of sharing are especially promising when sharing today is not only characterized by practices of sharing within one's inner circle of family, friends or neighbors, but increasingly also by practices of sharing with strangers who share little in common with each other (Frenken and Schor 2017).

This practice of sharing with strangers has been most systematically capitalized by the sharing economy. Despite the different and evolving practices that make up the sharing economy today, these practices can nonetheless be characterized by four key attributes. They are the recirculation of goods, an increase in the utilization of durable assets, exchange of services, and the sharing of productive assets (Schor 2014). Underlying these attributes is a commitment for sustainable and collaborative consumption, where instead of owning assets, individuals offer each other access to underused or idle assets (Botsman and Rogers 2011). But because sharing platforms have the tendency to scale up and dominate (Schor 2014), and because much of these platforms have been shaped by shared consumption at the expense of neglecting other social possibilities of sharing, the sharing economy has rapidly corporatized to provide just that: the consumption of shared assets in notable categories of mobility, workspace, and housing. By one estimate, the sharing economy is anticipated to grow to an estimated global revenue of nearly $335 billion (US dollars) by 2025 (PricewaterhouseCoopers 2015). It is no longer clear if the sharing economy is truly about sustainable or equitable practices anymore (see Schor 2017).

There are a number of issues and problems associated with the sharing economy, which we will further explicate in Chap. 2. But here, it suffices to say that the sharing economy has yet to tap into the prosocial and capacities-building possibilities of sharing in cities (McLaren and Agyeman 2015). In many cities, a rapidly expanding

J. K. H. Chan and Y. Zhang, *Sharing by Design*, SpringerBriefs in Applied Sciences and Technology, https://doi.org/10.1007/978-3-030-43569-1_1

urban population coupled with the growing scarcity of urban resources suggests that people would have to either share more, or else to increasingly compete for more scarce resources in the near future (see Huron 2015). But if relentless competition among people in the city is deemed undesirable, yet if the sharing economy remains unready to advance beyond the frame of corporatization and profitability, then there is little recourse but to seek out alternatives beyond the marketized notion of sharing. In tandem, the sharing economy remains largely devoid of the moral sentiments that could catalyze, drive, or sustain sharing practices. As Smith (1974) acutely observed long ago, these moral sentiments are necessary to restrain the excesses of self-interest. Anyone who had to hobble around shared bikes or scooters scattered across a sidewalk immediately understands that the maximization of self-interest even by shared consumption, independent of the responsibilities for a shared environment, is fundamentally self-defeating or missing the point of sharing altogether.

To further explore the prosocial and capacities-building possibilities of sharing, it is important to move away from the sharing economy by defining sharing anew. This does not mean that our argument denies the influential and impactful realities of the sharing economy today. On the contrary, our aim in this book is to provide a reformulation of sharing not only because of, but also in spite of, the sharing economy. In other words, the rapid and lucrative onset of the sharing economy has not given societies the necessary room to rethink and develop a different approach to sharing practices that can better serve the collective good. Given the limitations of the sharing economy, we argue that it is time to at least rethink sharing again.

On this, we propose to rethink sharing as a category of design. Broadly stated, we are interested in how artificial configurations of sharing with strangers can emerge and how we can further intervene to enhance or sustain them. Here, consider the iconic 'Oxford library of things', which is an actual tool library packed with many useful items for sharing (Benedictus 2019). Using this example, we may begin by asking the following design-oriented questions: How did this sharing practice come about? Who participates in sharing here? What are the conditions that facilitate sharing and what are the challenges faced by this sharing practice? And how can this practice be designed—materially, spatially, or organizationally—to further enhance its original mission? Finally, how should this sharing practice attract a future generation to continue to operate this tool library (see Bateson 2000: 512)—indeed, how should one design this system in order to secure improvements that can persist over time (Ulrich 1994)? Rethinking sharing as a category of design then changes the kind of questions that could be asked about sharing. Questions pertaining to the design histories, politics, social contexts, and ethics of sharing become salient because answers to these questions matter for the integrity and sustainability of the designed sharing system.

All these considerations can be summed up by the following two organizing questions: How exactly do we design sharing systems in cities today? And what is the content of the design knowledge for doing this? These two questions underpin our design research on sharing and we seek to answer them in this book. At least by the frame of design, it may become clearer that the sharing economy is only one of the many possible approaches to organize and materialize the sharing relation. And through the frame of design, it may become possible to conceive a far wider range of sharing configurations that then can be variably specified to the different contexts that require them. Relying on design, it may even be possible to influence, project, and propagate sharing practices on relations and spaces in the city that were once indifferent to sharing (see Katrini 2018).

Here, it is important to justify the frame of design. Like *homo sapiens,* many animals share resources. But no known animal is able to approach the scale and complexity that characterize human sharing (Davies 2019: 2). We are often impelled to continue to share when someone has shared with us (Belk 2007: 133), just as we create a fairly complex system like the tool library in order to share more effectively or efficiently. *Homo sapiens* may be the only species that is uniquely capable of synchronous sharing and designing. In other words, humans are capable of what Tomasello (2014: 4) calls, "shared intentionality": the ability to form a joint goal by collaboration. For example, we share tools or resources, or else communicate ideas and knowledge among other know-how in collaborative design in order to improve the performance or quality of the outcome. Because of this sharing, we also learn from each other. And learning in the design process then deepens and transforms the original design problem, which Schön (1990: 128) describes as how, "running the maze changes the maze". In turn, a transformed design problem then permits new possibilities and horizons that were latent before.

We find this unique human capability promising when applied to the domain of sharing. To describe this capability in a slightly different way, we are the only species capable of changing sharing configurations when they cease to work for us; we are also the only species that create sharing configurations anew when we perceive a need for them. Furthermore, we are able to deliberate among ourselves and then project a better configuration that is more ideal in some ways: configurations that perform sharing more effectively or efficiently, or else sharing is practiced in more equitable or beautiful ways. And when these criteria are no longer appropriate, we are able to discuss and seek out new criteria, which in turn invite proposals for more advanced sharing configurations that can better reflect our new ethical consensus.

This is then, the central idea of *sharing by design.*

1.1 What Exactly Is Design?

To further unpack the idea of sharing by design, it is important to first break down this phrase by first reviewing the concept of design, and then, sharing. To begin, no scholarly consensus exists yet on a unified definition of 'design'. Even so, it is possible to suggest that design presents a theory for change—especially a change for betterment (Krippendorf 2006). Simon (1996: 111), for instance, has defined design as devising "courses of actions aimed at changing existing situations into preferred ones". And more specifically, Rittel defined design as an activity that aims at the production of a plan, which—if implemented—is intended to bring about a situation with the desired consequences without creating unforeseen and undesired side and after-effects (Protzen and Harris 2010: 1). On the other hand, Schön (1990: 110) defined design as "ambiguously signifying process and product", entailing the entire range of artifacts made by human beings. Importantly because so much of the artificial world has come about because of design, Schön (1990) also tried to specify what is not design. For Schön (1990), random or purposeless behavior, processes that are completely proceduralized, and actions in which deliberate intervention plays negligible parts are not to be considered design. Design is quintessentially for Schön (1990: 139), "a conversation with the materials of a situation".

More recently, Murphy (2016) summarizes that design could be characterized by features of form, order, planning, and intention. Based on this, design could be defined as the purposeful act of planning for changing existing form or order in order to attain certain goals. And reorienting design away from its functionalist, rationalist, and industrial traditions, Escobar (2017) asks if design can be rerouted to materialize new relational dimensions of life. Importantly for Escobar (2017), design is about creating new cultural meanings, practices, and particular ways of living in a world that is shifting toward social fragmentation and environmental collapse. Design is one central activity through which a more sustainable culture can be given its material form (see Balsamo 2011: 196).

Even so, it is important to acknowledge that design—in spite of the range of its avowed and unambiguously constructive intentions—often misses its mark in reality. Rittel once referred to this as a form of design failure (Protzen and Harris 2010). A 'type-1' failure occurs if the design does not accomplish what was intended, while a 'type-2' failure occurs when the design leads to side and after-effects that were unforeseen and unintended, and further, proves to be undesirable as well. Rittel's typology of design failure is merely suggestive of a common reality in design practice. To evaluate if a design outcome has accomplished what was intended depends not only on the performance of this design, but also on whose intentions count as well. Similarly, what is considered undesirable will depend on what or whose ethics is involved in judging the design outcome. To improve the odds of effective design then, inclusivity or fair participation, and some degree of ethical consensus, are key considerations in the design of any large-scale sharing system.

1.2 The Many Meanings and Significance of Sharing Today

'Sharing' is homonymic (John 2017): this means that the actual meaning of 'sharing' changes in different context. Tracing the philological history of 'sharing', John (2017) outlines the range of meanings implied by this word. First, sharing is a way of dividing or distributing resources; this is also an early meaning of sharing. Second, sharing can imply the state of having something common with someone. Third, sharing can be an act of communication, experienced for instance, when we share our feelings or emotions. Fourth and more recently, sharing has been used to exemplify the return to a more morally superior and natural state of being. On this, John (2017) suggests that sharing represents an alternative to capitalism, yet at the same time—ironically through the sharing economy—it is also the mode of participation at capitalism's cutting edge.

And extending this list, sharing can be distinguished from gift-giving and commodity exchange (Belk 2010). On this, sharing is a practice—it is what people do (Barnett 2018). Furthermore, it is possible to distinguish between autotelic sharing and telic sharing (see Widlok 2004: 61). In the former, sharing is its own end; sharing is practiced for the sake of the shared enjoyment that it can confer. Conversely in the latter, sharing is practiced as an instrumental means to meet other objectives. And developing from telic, or purposeful sharing, it is possible to further distinguish between direct and indirect sharing. While direct sharing focuses on the resource that is intended to be shared, indirect sharing highlights the background conditions that are shared in order for direct sharing to occur. For instance, the direct sharing of shared bikes has to presume the indirect sharing of the urban environment in which sharing bikes takes place (Chan and Zhang 2018). More recently, against the neoliberal realities of privatization and enclosures in cities today, sharing has also been seen as a form of emancipatory politics: sharing practices not only generate spaces of economic alterity but also facilitates relations of connectedness and interrelationality (Hall and Ince 2018). Sharing accommodation, for instance, provides dignified shelter and amenities to those who otherwise have to settle for precarious dwellings (Bhatia and Steinmuller 2018). Other similar attempts have defined sharing as one important cooperative approach to address people's everyday needs by co-managing local resources (Katrini 2018; Petrescu et al. 2016).

This breadth of meanings suggests that there is no single and privileged ontology of sharing. Instead, the context shapes the meaning of sharing. For the purpose of this book then, it is important to circumscribe what sharing is, and what it is not in relation to design. Following this, we will also state what is excluded in our study on 'sharing by design'.

First of all, we specify what sharing is by defining it in contrast to what it is not in the context of sharing by design. By 'sharing', we are not referring to the state of having something in common by default—for instance, that we share the same world, or that in the same society, we have certain shared norms. In contradistinction, when we refer to 'sharing', we are referring to the agency of direct sharing: that we share some resources deliberately in order to satisfy everyone's needs, and that sharing is intentional—deliberately selected by individuals among all other modalities of distributing certain resources—in order to achieve a certain goal. In this usage of direct sharing, we have in mind Widlok's (2017: 111) characterization of sharing as a moral skill, rather than a technical capacity, that is used for the creation of a shared community. Following this, 'sharing' here also does not exclusively connote the adjectival use of 'share' as in the example of 'a shared resource', but instead and importantly for our thesis, refer to the action verb of 'share'—such as the relation, or process, underlying the example of 'sharing resources'. And finally, the thesis of sharing by design does not immediately rule out the more familiar idea of sharing within an inner circle of families, friends, or neighbors. However, our thesis is more vested in outlining the significance, mechanisms, and possibilities of sharing with strangers who share little in common. In sum, sharing by design is the thesis of a combined understanding of sharing as a form of agency, as a relational procedure, and as the configurational possibility of sharing with strangers.

At this point, we also state what we have excluded from our study on 'sharing by design'. Notably, we have excluded the techno-cultural phenomenon of data sharing and the moral dilemmas that are surfaced by this sharing practice. While we remain deeply vested in the ethical dimensions of sharing and have relied on an actual case of video sharing to prime our discussion on ethics and sharing in Chap. 5, however, we suggest that the practice of data sharing belongs to a broader analysis of digital culture, which, we might also add, the warrant of a separate book-length project. For this reason, we have decided to exclude a more thorough engagement with this significant but contentious data sharing practice in our focused discussion on sharing by design. By specifying what is excluded, we suggest that this will render our focus and emphasis on sharing by design not only sharper but also much clearer.

1.3 Why Is There a Need for Sharing by Design?

In spite of this breadth of meanings, we usually do not have trouble knowing what to share or what not to share, how or when to share certain information or things with others in our everyday life. In Goffman's (2005) words, we all have some knowledge of tact, *savoir-faire* or social skills in these instances of inter-personal sharing. We may share too much or too little—willingly or unwillingly—but rarely would we not know how to share. Much of such inter-personal sharing is also constrained, or otherwise prompted by social norms and cultural expectations. After all, sharing is expressed in ethical systems, in religions and in many social forms and rituals (Price 1975: 6), and evidence suggests that sharing is a culturally learned behavior (Belk 2007: 130). Even young children know when to share—especially when another child is watching (Tomasello 2014: 47).

Challenges emerge, however, when sharing takes place within environments that have been artificially organized to foster sharing relations. In other words, these are environments that have been deliberately conditioned, or operationally structured, for sharing to take place. For instance, in one landmark study on the co-working space, Merkel (2017) notes that while this new type of workplaces offers co-location and also co-presence for workers—which are preconditions for many forms of sharing relations—it remains unclear what and how co-workers share when they have different professional interests and knowledge backgrounds. In another study on hackerspaces, Davies (2017) observes that the democratic nature of this shared space ironically renders sharing governance responsibility challenging. These challenges are also perceived in co-housing (Sargisson 2018), where shared ownership of this model requires joint decision-making on how to share space, which in turn increases interaction between neighbors. But this increased interaction does not automatically produce harmonious relations. For this reason, Sargisson (2018) argues that some form of social design is also necessary in co-housing. In sum, these challenges suggest that nascent models of sharing by design already exist, but they are likely to require further tinkering by design to improve sharing behaviors and relations.

Beyond these challenges, there are at least three other reasons for why we may want to consider sharing by design. First, the fact that individuals participate in the social and socializing process of sharing neither presupposes sociality and positive social relations, nor does the act of sharing automatically align to a more just redistribution of resources (Davies and Evans 2018). In other words, sharing does not presume equity. Sharing can in fact worsen inequality (McLaren and Agyeman 2015; Schor and Attwood-Charles 2017). To guarantee equity, it is not enough just to ensure access to shared resources. Importantly, synthetic configurations in the form of governance structure, guiding principles, and norms have to be designed to ensure equity in the access and use of the shared resource (see Ostrom 2006). To formulate this slightly differently, to commit to *laissez faire* in sharing is likely to result in the abuse of shared resources. Some form of regulatory interventions and re-alignments are necessary to ensure sustained and fair sharing.

Second, sharing has been observed as an emerging urban phenomenon (McLaren and Agyeman 2015)—not least because the population density of cities tends to promote an 'economy of scale' in sharing resources (Rose 1986), but also that cities today constitute the background of a struggle for increasingly scarce resources that sharing may come to mitigate (Hall and Ince 2018). But integrating a new sharing practice into an existing urban neighborhood, structured with its own spatial and administrative politics, is never without antagonism. For example, coping with the power relations of the municipal administration has been perceived as a key difficulty of maintaining even a robustly managed sharing community (Petrescu et al. 2016). Furthermore, new sharing practices can impact their urban neighborhood (Katrini 2019). In one study, researchers argue that diffusing co-working spaces in Milan may have the urban effects of positively enhancing community, improving surrounding public spaces, and revitalizing the city (Mariotti et al. 2017: 57). But by the same token, new sharing practices can also increase rents, bring about negative externalities (for instance, noise and traffic), and lead to evictions and displacements. In other words, how a sharing practice integrates into the urban neighborhood with as little negative impact as possible is a design consideration external to the constitution of the sharing practice itself. To ensure a sustained sharing practice in any urban context, this practice has to be designed.

Finally, for certain sharing practices to be even possible, there is a need to plan for sharing, which is just another way of framing sharing by design. Hult and Bradley (2017: 598) suggest that many forms of sharing practices require "sharing infrastructure"—defined as socio-technical infrastructure for sharing resources, tools, and skills. This infrastructure is important not only because it can facilitate sharing, but it can also enable new sharing practices. Drawing on the work of Rose (1986), Frischmann (2012) argues that sharing the same infrastructure increases participation in socially valuable activities, which yields scale returns downstream. If sharing is perceived as one of the truly sustainable ways forward for re-organizing present socio-economic models of production and consumption, then planning the necessary sharing infrastructure consolidates the preconditions of sharing by design.

1.4 An Overview of This Book

The rest of this book consists of working out the ideas discussed in this chapter. In Chap. 2, we will explicate the social traits of sharing in the sharing economy in relation to access, sharing with strangers, and digital platforms. In Chap. 3, we discuss the idea of 'sharing by design', and furthermore, describe two major paradigms that embody this idea today. Following closely in Chap. 4, we discuss how to design a sharing system. We begin by describing and then elaborating on the various components of a sharing system, before briefly simulating how to design a hypothetical waste-to-energy sharing system. Subsequently in Chap. 5, we examine sharing ethics through the ethical categories one is likely to encounter in sharing relations and practices, namely, virtues, principles of justice and beneficence, and ethical expertise. Following this, in Chap. 6, we discuss the pedagogy of sharing by design by reflecting on how this was taught in an architectural and urban design studio at the National University of Singapore. Finally, in Chap. 7, we reflect on our work and conclude on several important issues and questions that can point the way forward for future research and practice.

References

A. Balsamo, *Designing Culture: the Technological Imagination at Work* (Duke University Press, Durham, NC, 2011)

C. Barnett, Foreword, in *Sharing Economies in Times of Crisis: practices, Politics and Possibilities*, ed. by A. Ince, S.M. Hall (Routledge, New York, NY, 2018), pp. x–xiv

G. Bateson, *Steps to an Ecology of Mind* (University of Chicago Press, Chicago, 2000)

R. Belk, Why not share rather than own? Annal. Am. Acad. Polit. Soc. Sci. **611**(1), 126–140 (2007). https://doi.org/10.1177/0002716206298483

R. Belk, Sharing. J. Consum. Res. **36**(5), 715–734 (2010). https://doi.org/10.1086/612649

L. Benedictus, The library of things: could borrowing everything from drills to disco balls cut waste and save money? *The Guardian* (2019), https://www.theguardian.com/society/2019/apr/24/library-of-things-borrowing-scheme-conquer-world

N. Bhatia, A. Steinmuller, Spatial models for the domestic commons: communes, co-living and cooperatives. Archit. Des. Spec. Issue: Hous. Interv.: Archit. Towards Soc. Equity **88**(4), 120–127 (2018)

R. Botsman, R. Rogers, *What's Mine is Yours: the Rise of Collaborative Consumption* (Harper-Business, New York, NY, 2011)

J.K.H. Chan, Y. Zhang, Sharing space: urban sharing, sharing a living space, and shared social spaces. Space Cult. (2018). https://doi.org/10.1177/1206331218806160

S.R. Davies, *Hackerspaces: making the Maker Movement* (Polity, Malden, MA 2017)

A. Davies, D. Evans, Urban food sharing: emerging geographies of production, consumption and exchange. Geoforum (2018). https://doi.org/10.1016/j.geoforum.2018.11.015

A.R. Davies, *Urban Food Sharing: rules, Tools and Networks* (Policy Press, Bristol, UK, 2019)

A. Escobar, *Designs for the Pluriverse: radical Interdependence, Autonomy, and the Makings of Worlds* (Duke University Press, Durham, NC, 2017)

K. Frenken, J. Schor, Putting the sharing economy into perspective. Environ. Innov. Soc. Trans. **23**, 3–10 (2017)

B.M. Frischmann, *Infrastructure: the Social Value of Shared Resources* (Oxford University Press, Oxford, UK, 2012)

E. Goffman, *Interaction Ritual: essays in Face-to-Face Behavior* (Transaction Publishers, New Brunswick, NJ, 2005)

S.M. Hall, A. Ince, Introduction: sharing economies in times of crisis, in *Sharing Economies in Times of Crisis: practices, Politics and Possibilities*, ed. by A. Ince, S.M. Hall (Routledge, New York, NY, 2018), pp. 1–15

A. Hult, K. Bradley, Planning for sharing—Providing infrastructure for citizens to be makers and sharers. Plan. Theory Pract. **18**(4), 597–615 (2017)

A. Huron, Working with strangers in saturated space: reclaiming and maintaining the urban commons. Antipode **47**(4), 963–979 (2015)

H. Jarvis, Sharing, togetherness and intentional degrowth. Prog. Hum. Geogr. **43**(2), 256–275 (2019), https://doi.org/10.1177/0309132517746519

N.A. John, *The Age of Sharing* (Polity, Malden, MA, 2017)

E. Katrini, Sharing culture: on definitions, values, and emergence. Sociol. Rev. Monogr. **66**(2), 425–446 (2018)

E. Katrini, *Creating the Everyday Commons: spatial Patterns of Sharing Culture*. Unpublished Doctoral Dissertation (Carnegie Mellon University, Pittsburgh, PA, USA, 2019)

K. Krippendorf, *The Semantic Turn: a New Foundation for Design* (CRC Press, Boca Raton, FL, 2006)

I. Mariotti, C. Pacchi, S. Di Vita, Co-working spaces in Milan: location patterns and urban effects. J. Urban Technol. **24**(3), 47–66 (2017)

D. McLaren, J. Agyeman, *Sharing Cities: a Case for Truly Smart and Sustainable Cities* (MIT Press, Cambridge, MA, 2015)

J. Merkel, Coworking and innovation, in *The Elgar Companion to Innovation and Knowledge Creation*, ed. by H. Bathelt, P. Cohendet, S. Henn, L. Simon (Edward Elgar, Cheltenham, UK, 2017), pp. 570–586

K.M. Murphy, Design and anthropology. Annu. Rev. Anthropol. **45**, 433–449 (2016)

E. Ostrom, *Governing the Commons: the Evolution of Institutions for Collective Action* (Cambridge University Press, New York, NY, 2006)

D. Petrescu, C. Petcou, C. Baibarac, Co-producing commons-based resilience: lessons from R-urban. Build. Res. Inf. **44**(7), 717–736 (2016)

J.A. Price, Sharing: the integration of intimate economies. Anthropologica **17**(1), 3–27 (1975)

PricewaterhouseCoopers, *The Sharing Economy* (2015), https://www.pwc.com/us/en/services/consulting/library/consumer-intelligence-series/sharing-economy.html

J.P. Protzen, D.J. Harris, *The Universe of Design: Horst Rittel's Theories of Design and Planning* (Routledge, New York, NY, 2010)

C.M. Rose, The comedy of the commons: commerce, custom, and inherently public property. Univ. Chic. Law Rev. **53**(3), 711–781 (1986)

L. Sargisson, Swimming against the tide: collaborative housing and practices of sharing, in *Sharing Economies in Times of Crisis: practices, Politics and Possibilities*, ed. by A. Ince, S.M. Hall (Routledge, London, UK, 2018), pp. 145–149

D.A. Schön, The design process, in *Varieties of Thinking*, ed. by V.A. Howard (Routledge, New York, NY, 1990), pp. 110–141

J.B. Schor, W. Attwood-Charles, The "sharing" economy: Labor, inequality, and social connection on for-profit platforms. Sociol. Compass (2017), https://doi.org/10.1111/soc4.12493

J.B. Schor, *Debating the Sharing Economy* (2014), https://greattransition.org/publication/debating-the-sharing-economy

J.B. Schor, Does the sharing economy increase inequality within the eighty percent?: findings from a qualitative study of platform providers. Camb. J. Reg. Econ. Soc. **10**, 263–279 (2017)
H.A. Simon, *The Sciences of the Artificial*, 3rd edn. (MIT Press, Cambridge, MA, 1996)
A. Smith, *The Theory of Moral Sentiments* (Penguin Books, New York, NY, 1974)
M. Tomasello, *A natural History of Human Thinking* (Harvard University Press, Cambridge, MA, 2014)
W. Ulrich, Can we secure future-responsive management through systems thinking and design? Interfaces **24**(4), 26–27 (1994)
T. Widlok, Sharing by default? outline of an anthropology of virtue. Anthropol. Theory **4**(1), 53–70 (2004)
T. Widlok, *Anthropology and the Economy of Sharing* (Routledge, New York, NY, 2017)

Chapter 2
Assessing the State of Sharing in the Sharing Economy

2.1 Sharing as a Social Act

Sharing is generally defined as using, occupying, enjoying, and having something jointly with others; possessing some qualities in common with others; and communicating to others about something (Oxford Dictionary). It is often understood, from different perspectives, as an evolutionary trait of human beings (Boehm 1999), a type of consumption behavior (Belk 2010), a way to maintain intimate interpersonal relations, more recently an emerging form of economic activities (Botsman and Rogers 2010) and an act of participation in the age of social media platforms (John 2013a). Despite the various prisms through which sharing has been understood and investigated, sharing essentially has a deep social root and is always constitutive of certain social relations. Before the emergence and prevalence of digital technologies, sharing mostly took place in the physical world, and irrespective of the motivations and purposes, always increased the intensity of social interactions and social exchange (Wittel 2011). In an increasingly digitalized era, despite that sharing is often criticized for being used as a metaphor and a quasi-technical term for digital data transfer and associated wealth accumulation (John 2013b), it still has great potential to trigger social responses, create social interactions, and strengthen social ties (e.g. Belk 2014a; Wittel 2011). From this perspective, it is not only necessary but also crucial to explicate the key social traits that define sharing, in order to better understand sharing in the context of the sharing economy.

First, in stark contrast to calculative exchange, sharing is defined as a prosocial and non-reciprocal behavior (Benkler 2004). As the old saying goes, sharing is caring. The motivation of sharing is not to pursue self-interest, but instead to help and give without reciprocal expectations (Belk 2007). Conventionally, this altruistic nature of sharing can be best exemplified by a mother freely giving her nurturing, care, and love to the infant without tracking the balance between giving and receiving (Belk 2010). The same can also be observed in the household pooling and sharing of resources, whereby each member contributes according to his or her abilities to meet the needs of the family as a whole (Thomas 1990). The moral principles of sharing,

J. K. H. Chan and Y. Zhang, *Sharing by Design*, SpringerBriefs in Applied Sciences and Technology, https://doi.org/10.1007/978-3-030-43569-1_2

such as empathy, fairness, selflessness, and equality, are often found embodied in the ubiquitous sharing practice today, which is regarded both as an act of communication (e.g. online sharing of files and photos) and a type of distribution (e.g. car sharing or tool library as in the sharing economy) (John 2013b, 2017). For instance, many members of GitHub who openly share their codes of computing models are not hoping to benefit from others' sharing. Rather, their priority is to help those who are in need and to contribute to advancing the model development. Not limited to online activities, in the public housing of Singapore, two full-sized refrigerators were installed in the public area for residents to share various food, even including fresh catches from the sea, with their needy neighbors and especially the poor and the marginalized (Lin 2017).

Second, sharing often entails a joint possession of what is being shared, even if only temporarily. Because of this, sharing usually defines something as ours instead of distinguishing mine from yours (Belk 2007). This means that while people involved in sharing can freely enjoy the benefits that flow from possessing something together, at the same time they also have to take the responsibilities for sustaining the shared as well as the sharing practice. For instance, one may gain close friendship with a flat mate from sharing a living space, but this is nonetheless largely predicated on the fact that one has to take care of the shared environment without causing any detriments to the everyday life of each other. In other words, it is the joint possession and especially the concomitant responsibilities that lead to form a sense of togetherness (Jarvis 2019). Because of the joint possession, sharing clearly contrasts with both gift-giving and market exchange, of which both always necessitate the transfer of ownerships and also differentiate givers from receivers, and sellers from buyers (Belk 2010). It is important to note the nuance between joint possession and collaborative ownership, of which the latter often technically enables people to 'share' the cost (bear the cost together) and then gain access to what they could not have afforded on their own. However, the accessed goods and services are rarely truly shared, let alone encouraging individuals to take any associated responsibilities.

Third, sharing is often a powerful communal act that connects individuals, creates trust and bonding, and engenders a sense of community (Belk 2010). Whenever one shares altruistically, whether with family members, friends, neighbors, strangers, or even the unknown, he or she is in fact making a great communitarian commitment and contributes to community building. This is because one's generosity embodied in sharing is likely to encourage and motivate others who benefit from what is being shared to do the same. Such mutual influence and the resultant continuous sharing process will then help to develop and enhance individuals' sense of community, which in turn may further foster and reinforce the sharing practice. Sharing reinforced by this virtuous circle is arguably crucial for the development of human societies. Studies of the hunter–gatherer societies reveal that the non-utilitarian and equal sharing was always considered central to resource efficiency, security, and most importantly, survival of tribes (e.g. Price 1975; Stack 1974; Widlok 2004). From the anthropological perspective, Fiske (1991) regarded sharing as one of the four basic types of social relationships that creates unity and solidarity. At present, no matter sharing online or in the physical world, although we are technically sharing different goods and services such as music, files, information, beers, books, and many others, the true values that are anticipated from the sharing practices are social connection, mutual trust, bonding capital, friendship, and community (e.g. Steinkuehler and Williams 2006).

These three social traits of sharing are not independent from each other. As indicated above, the altruistic motivation of sharing is often the basis for trust development and community building. In turn, the benefits from communities are likely to further encourage sharing and reinforce one's responsibility for sustaining sharing. Taken together, these traits are encapsulated in what Belk (1988) referred to as 'extended self'. Depending on specific contexts, this notion may expand from within the family, which is the most immediate level of extended self, to including strangers or even unseen others, which is generally understood as the aggregate extended self. Within the circle of extended self, one no longer emphasizes the boundaries that separate self and others and the difference between givers and receivers. This means that sharing with others is like sharing with self; therefore, it is usually altruistic and non-reciprocal, encouraging voluntary responsibilities, and contributing to creating a sense of mutuality and community. Belk (2010) regarded the sharing within the sphere of extended self as 'sharing in', which is in contrast with 'sharing out' that tends to preserve the self/other boundaries and is usually driven by self-interest and closer to commodity exchange. 'Sharing out' may sometimes technically involve joint possession of what is being shared, but because the sharers are not treated as being within the circle of extended self, one is unlikely to make a communitarian commitment. 'Sharing out' thus often lacks a communal dimension.

In the remaining of this chapter, we will take the three traits of sharing as well as the notion of 'sharing in' vis-à-vis 'sharing out' as a prism to examine to what extent and in what ways sharing, especially its social qualities, have been incorporated in the prevalent sharing economy today.

2.2 Is Sharing Fully Reflected in Framing the Sharing Economy?

The sharing economy is generally regarded disruptive to the mainstream market and is claimed to be economically fairer, environmentally more sustainable, and socially more connected (e.g. Botsman and Rogers 2010). With a great diversity of activities parked under this big tent, different terms have been coined to describe and define the sharing economy, but each usually cuts in from a specific angle. So far, there is still a lack of consensus on a solid definition of this new phenomenon, and a unitary definition may be nearly impossible according to Schor (2016).

To better gain a clearer understanding of how sharing has been reflected in the framing of the sharing economy, we have incorporated in our analysis the most commonly used terms associated with the sharing economy (summarized in Table 2.1). Through this enumerating methodology, it is then possible to review the details

Table 2.1 A summary of the most commonly used terms in the discourse of the sharing economy

Sharing economy	"An economic model based on sharing underutilized assets from spaces to skills to stuff for monetary or non-monetary benefits." (Botsman 2013) "Coming up with a solid definition of the sharing economy that reflects common usage is nearly impossible…. Sharing economy activities fall into four broad categories: recirculation of goods, increased utilization of durable assets, exchange of services, and sharing of productive assets." (Schor 2016: 2) "The sharing of under-used assets through completing peer-to-peer transactions that are only viable through digital intermediation, allowing parties to benefit from usage outside of the primary use of that asset." (Beck et al. 2017: 3) "consumers granting each other temporary access to under-utilized physical assets ("idle capacity"), possibly for money. (Frenken and Schor 2017: 4–5)
Collaborative economy	"An economy built on distributed networks of connected individuals and communities versus centralized institutions, transforming how we can produce, consume, finance, and learn. It has four key components: Production: Design, production, and distribution of goods through collaborative networks Consumption: Maximum utilization of assets through efficient models of redistribution and shared access Finance: Person-to-person banking and crowd-driven investment models that decentralize finance Education: Open education and person-to-person learning models that democratize education." (Botsman 2013) "An economic model where ownership and access are shared between corporations, start-ups, and people. This results in market efficiencies that bear new products, services, and business growth" (Owyang et al. 2013: 4) "…using internet technologies to connect distributed groups of people to make better use of goods, skills and other useful things…. (it has) five collaborative economy traits: 1. Enabled by internet technologies. 2. Connecting distributed networks of people and/or assets. 3. Making use of the idling capacity of tangible and intangible assets. 4. Encouraging meaningful interactions and trust. 5. Embracing openness, inclusivity and the commons." (Stokes et al. 2014: 10–12)

(continued)

Table 2.1 (continued)

Collaborative consumption	"Those events in which one or more persons consume economic goods or services in the process of engaging in joint activities with one or more others." (Felson and Spaeth 1978: 164) "An economic model based on sharing, swapping, trading, or renting products and services, enabling access over ownership. It is reinventing not just what we consume but how we consume. It has three distinct systems: 1. Redistribution markets: Unwanted or underused goods redistributed 2. Collaborative lifestyles: Non-product assets such as space, skills, and money are exchanged and traded in new ways 3. Product service systems: Pay to access the benefit of a product versus needing to own it outright." (Botsman 2013) "people coordinating the acquisition and distribution of a resource for a fee or other compensation. By including other compensation, the definition also encompasses bartering, trading, and swapping, which involve giving and receiving non-monetary compensation. But this definition of collaborative consumption excludes sharing activities like those of CouchSurfing because there is no compensation involved…. The definition also excludes gift giving which involves a permanent transfer of ownership." (Belk 2014b) "Market model that enables individuals to coordinate the acquisition and distribution of a resource for a fee or other compensation (Belk 2014b), where the interaction is at least partially supported or mediated by technology." (Perren and Grauerholz 2015: 141) "The peer-to-peer-based activity of obtaining, giving, or sharing access to goods and services, coordinated through community-based online service. CC has been expected to alleviate societal problems such as hyper-consumption, pollution, and poverty by lowering the cost of economic coordination within communities." (Hamari et al. 2016: 2047)
Mesh	"Mesh businesses … use data crunched from every available source to deliver high-quality goods and services to people only when they need and want them. Mesh businesses share four characteristics: sharing, advance use of web and mobile information networks, a focus on physical goods and materials, and engagement with customers through social networks." (Gansky 2010: 15–16) "The mesh, a term coined by author and investor Lisa Gansky in 2010, alludes to the way in which digital technology is used to provide people with access to goods and services, as they want them, in new and interesting ways." (Stokes et al. 2014: 10) "This is a type of network that allows any node to link in any direction with any other nodes in the system. Hence, it is about the sharing or meshing of talents, goods and services, in which the reference model is based on a series of transactions, on sharing something over and over… multiple sales multiply profits and customer contact, and multiple contacts multiply opportunities for additional sales and for deepening and extending relationships with customers." (Selloni 2017: 17)

(continued)

Table 2.1 (continued)

Peer-to-peer economy	"Person-to-person marketplaces that facilitate the sharing and direct trade of assets built on peer trust." (Botsman 2013) "organizations built around a peer–to–peer business model, whereby people use platforms to rent, sell, lend or share things with others without the involvement of shops, banks or agencies." (Stokes et al. 2014: 9) An economy that concerns goods that are also under-used but which are offered directly by their owners. (Pais and Provasi 2015 extracted from Selloni 2017: 18)
Access-based consumption	"Transactions that can be market mediated but where no transfer of ownership takes place." (Bardhi and Eckhardt 2012: 881) "market-mediated transactions that provide customers with temporarily limited access to goods in return for an access fee, while the legal ownership remains with the service provider" (Schaefers et al. 2016: 3). "consumers are paying to access someone else's goods or services for a particular period of time. It is an economic exchange, and consumers are after utilitarian, rather than social, value." (Eckhardt and Bardhi 2015)
Gig economy	"An important subset of the collaborative economy: the new platforms that provide flexible work such as errand marketplace TaskRabbit and on-demand ridesharing platform Lyft." (Stokes et al. 2014: 10) "the repercussions of the sharing economy on the labour market, where increasing numbers of people carry out successive jobs via the platforms which have been developed to act as intermediaries between individuals for the purpose of those activities, but which do not actually have any direct employees." (Basselier et al. 2018: 58)

Source Authors

afforded by each term at one level, and on another level, to behold all these terms in one single view.

It is evident from the various terms that the sharing economy is primarily framed as a new economic opportunity. Most framings revolve around the proposition that the sharing economy creates new economic opportunities, including but not limited to renting, swapping, trading, exchanging, giving, and of course, sharing. These opportunities then lead to more efficient distribution of resources, especially the underutilized and idled resources. In this way, the sharing economy as a disruptive innovation transforms the market model and encourages decentralized production and consumption. The dominance of this perspective seems to suggest that the claimed environmental and social sustainability of the sharing economy is largely secondary to, if not entirely dependent on, its economic achievements. Martin (2016) confirmed the superior position of these economic opportunities across different framings of the sharing economy, which range from a disruptive transition leading to great sustainability at one end, to a disguised form of neoliberalism at the other. Nevertheless, despite the hitherto lack of emphasis on the social values, the very notion of sharing is still central to this new phenomenon.

The word 'sharing' can be easily identified in the descriptions of almost every single term listed in the table. It is generally perceived as a consumption behavior that usually seeks benefits or compensations. Sharing is, for example, to 'benefit from usage outside of the primary use of that asset' in (Beck et al. 2017); to achieve 'market efficiencies that bear new products, services, and business growth' in (Owyang et al. 2013); and for 'monetary or non-monetary benefits' in (Botsman 2013) and 'possibly for money' in (Frenken and Schor 2017). This is not to say that there exists no genuine and altruistic sharing in the sharing economy, nor that the moral motivation of sharing such as caring and helping are completely irrelevant. It is not unfamiliar for many that some Couchsurfing hosts do share their homes with the intent of enhancing the travelers' experience. Similarly, the redistribution market, which is a key category of the sharing economy according to Botsman and Rogers (2010), does allow many to share (i.e. to give without any expectation of compensation) their unwanted and underused goods with those who have genuine needs. These seem to suggest that motivations of individuals' participation in the sharing economy are likely to be mixed. This then leads to the question about the relation between different intents of sharing. Would the benefit-seeking purpose suppress, and even undermine, the altruistic and communal pursuit of sharing?

More specifically, 'sharing' is mostly interpreted in the framings as shared access over ownership to goods and services. This is also deemed as the key element of how the sharing economy transforms the ways we produce and consume. Shared access can be gained through, for instance, collaborative network (Botsman 2013), connecting distributed groups of people (Stoke et al. 2014), coordinated acquisition (Belk 2014b), peer-to-peer-based exchanges, and transactions (Hamari et al. 2016; Perren and Grauerholz 2015). While some approaches such as selling and exchange entail transfer of ownership, none of them by itself requires nor necessitates joint possession of whatever it is that is being shared. This means that individuals are very likely to solely enjoy the benefit of accessing to a rich variety of resources that they would otherwise unable to afford, but without being obliged to bear the responsibilities for maintaining the shared and sustaining this sharing practice. This is fundamentally different from perceiving sharing primarily as a social act—accentuating shared access greatly weakens and dilutes the notion of 'ours' in the sharing practice and consequently sharing is mostly likely to be 'sharing out' and less communal. It is clear from the table that both the idea of joint possession and the importance of sustaining the sharing and the shared are not incorporated in any single framing of the sharing economy. This begs another question about the continuity of sharing and the likelihood of trust building and community formation.

In addition, it is also noticeable from the table that underpinning the shared access is an unprecedented network of individuals and communities connected by digital platforms. The social benefits of this network for sharing, such as more opportunities for interactions and higher chances of bonds and trust formation, are acknowledged in many framings, though not fully explicit. For instance, Stock et al. (2014) explicitly emphasize meaningful interactions and trust as a key value. Harami et al. (2016) also suggest the contributions to alleviating many problems through facilitating community-based coordination. Nevertheless, the economic value of the distributed network is still greatly prioritized in the framings—that is, the network is mainly perceived as enabling and fostering shared access for benefits and compensations. This seems to suggest the unleashed potential of an extensive network that can facilitate sharing and foster social bonding.

2.3 Is Sharing Fully Practiced in the Sharing Economy?

From above, we can see that sharing as a social act is only partially reflected in the framings of the sharing economy. To use Belk's (2010) words, sharing in the sharing economy seems to be perceived very close to 'sharing out', whereby the self and others boundaries remain strong and the range of extended self is limited. It has to be clarified, however, that our discussion is not meant to debunk the sharing economy, nor does it intend to deny its many values that have been widely discussed elsewhere. Rather, our analysis suggests that the sharing economy is in fact multifaceted and that sharing in the context of the sharing economy, though primarily framed as a consumption behavior, may have different possibilities. In the following, we will continue with the analysis by looking into sharing in the practice of the sharing economy at three key dimensions, namely access, sharing with strangers, and digital platforms. We will revisit our focus on the three social traits of sharing raised earlier in this chapter—prosocial and altruistic motivation; joint possession and associated responsibility; bonds formation and community building—in examining the following dimensions of the sharing economy.

2.3.1 Access

As mentioned above, sharing is generally framed in the sharing economy as shared access over ownership to a great variety of goods and services. What drives individuals' pursuit of access are firstly and primarily self-interest and utilitarianism (Bardhi and Eckhardt 2012). For those seeking access, the economic consciousness of saving money is always intuitive; for those providing access, monetizing their underutilized assets as well as their skills and knowledge usually stands at the core. Following this logic, sharing practice based on access is very likely to be impersonal and underscored by calculation and reciprocal expectation for each other. This is, however, not to say that caring and helping are completely absent from the intent of the access providers, nor that access seekers are not looking for social connections. Nevertheless, it is noteworthy that what influences one's attitude toward the sharing economy can be very different from what actually drives one's participation. Based on a thorough survey on the users of many sharing economy platforms, Hamari et al. (2016) pointed out that sustainability concerns, both social and environmental, are closely associated with people's perception of the sharing economy, but are not directly linked to their engagement. Instead, it is the economic benefits that have positive impact on individuals' intention to participate. Furthermore, Lawson et al. (2016) found that none of the key factors, including saving money, enhancing status, seeking variety, that can influence people's willingness to participate in the access-based consumption is socially oriented. These seem to suggest that while the sharing economy does not necessarily eliminate altruism and non-reciprocity, the predominant pursuit of access somehow may refrain people from practicing sharing with prosocial intents.

In addition to gaining economic benefits, forgoing property ownership and celebrating flexible access have increasingly become a popular means of identity construction and enhancement. In the sharing economy, access attains new symbolic capital related to economic viability, environmental sustainability, and social equality. Participating in the sharing economy thus becomes a strategy for many to improve their social standing and status (Lawson et al. 2016) and to express their ideological interests and practice their citizenship outside the marketplace (Bardhi and Eckhardt 2012). However, this dramatic shift in the politics of consumption can generate many externalities. One possible risk is that while the access to a wide range of resources does liberate consumers from many obligations and uncertainties that are usually associated with ownership (Schaefers et al. 2016), their emotional and social attachments to any properties are inevitably divorced at the same time. As a result, on the one hand, individuals may lose the needs for becoming attached to a brand community and connected to other like-minded owners. Instead, they would simply seek and float between a variety of resources without any bonding (Eckhardt and Bardhi 2015). On the other hand, access does not necessarily entail joint possession of properties, let alone the associated responsibility for sustaining the shared and the sharing. Access alone is, therefore, unlikely to foster a community or to create a commons. This then suggests a paradox of the sharing economy that access is simultaneously a constructive force that can bring together and link similar individuals, and a destructive force that can impede the formation of social bonds and communities.

Despite the primary goal of gaining economic and symbolic benefits, and despite the lack of prosocial driver of participation, access, especially when coupled with the extensive networks of distributed individuals, essentially provides a vast opportunity for building new social connections. However, access alone does not warrant the production of new relations and formation of communities. For instance, accessing to and staying at some one's domestic place may not necessarily yield new friendships (Sung et al. 2018), whereas participating in toy libraries (Ozanne and Ballantine 2010) and makerspaces (Davis 2017; William and Hall 2015) often gives rise to new social networks. Access platforms like the latter ones are usually not-for-profit and operating at the community level. In these cases, sharing (sharing access to goods and services) are very close to, if not fully the same as, a social act—participants are often motivated by social gains and willing to offer help while enjoying the material benefits of access; they tend to make communitarian commitment and show care and love for others and also the shared resources; they develop trust and bonding through interactions and pave the way to community building. Nevertheless, the market mediation renders such access platforms less social and more profit-driven. The fast-growing scale of individual platforms and anonymous transactions increasingly inhibit social interactions and bonding through access (Bardhi and Eckhardt 2012). Especially, from the access providers' point of view, they are also unlikely to hinge on the access platforms that can provide the most social benefits, but instead tend to shift their focuses to platforms offering the lower transaction cost. Moreover, Richardson (2015) highlights a larger threat of market mediation that the access platforms may expand from the private sector to replace, and to extract revenues from public resources, which are already scarce in expensive and densely populated urban areas.

2.3.2 Sharing with Strangers

Regardless of specific motivations and goals of participation, in the sharing economy individuals are anticipated to be connected to, and sharing with, a broad range of known and especially unknown others. Sharing with strangers beyond family and kin is also regarded as a key defining character of the sharing economy (Frenken and Schor 2017). Schor and Fitzmaurice (2015) further highlighted that the contemporary sharing economy is also distinguished by the participation of high cultural capital consumers who are electing to adopt this practice. Such a broad connection of individuals, especially among the unknown and across different social classes, seems to be promising for many social gains. Nevertheless, we need to look at how sharing is exactly practiced between strangers.

At the individual level, the social effects are dependent on the level of interaction required by the sharing practice as well as that desired by the participants. Individuals solely sharing access to whatever they cannot afford by themselves are usually unlikely to seek extensive interactions with each other, and for them, prosocial and altruistic motives of sharing can be trivial. Often, this kind of sharing activity such as car sharing is also not designed to necessitate interactions. Conversely, in sharing a collectively pooled and socially constructed common resource, individuals are more likely to be motivated by their altruism and social pursuits and eagerly desire for inter-personal interactions, even if the sharing practice has not been deliberately designed to do so. For instance, in their study of families' uses of toy libraries, Ozanne and Ozanne (2011) showed that while the functional benefits of gaining inexpensive access to a large variety of toys are their key considerations, equally crucial is their willingness to share experiences and to offer mutual supports through the informal social network formed based on the sharing of toys.

Nevertheless, the intentions of interacting with strangers are not always so straightforward. Motives behind individuals' desires for inter-personal interactions may not necessarily be about care and help, nor are they mainly to build new social relations. Take for example the accommodation sharing, where interactions between participants are most likely, if not fully essential. Regardless of the involvement of monetary transactions, one of the main reasons why a host shares a living space or the home with a stranger is the enjoyment gained from interacting with the unknown (Tussyadiah 2016). The host usually does not expect the relationships with guests would extend beyond that (Ikkala and Lampinen 2015). In turn, the same may also be said for the travelers who choose to stay in shared accommodations. This is certainly not to say that such sharing practices are socially not meaningful at all, nor does it suggest that the participants are fully driven by self-interest with little altruistic intents. The enjoyment of interactions, after all, comes from non-reciprocal sharing of knowledge and information. Nevertheless, this raises the question about the degree of the social impact of sharing with strangers. Research show that even though sharing of strangers' home may lead to the formation of new friendships (Parigi et al. 2013), they are unlikely to be durable (Parigi and State 2014).

At the community level, the formation of community in the age of the sharing economy is largely contingent upon individuals' participation in sharing practice and usually oriented around a particular type of shared resources or services (Richardson 2015). Enacted by digital platforms, sharing practice is neither predicated on spatial proximity, nor is it place-bound. This means that communities can emerge from within a neighborhood and primarily comprising of local residents or, more than often, expand across a region or even the whole world and connecting a great variety of people. For instance, a strong global community of Couchsurfers was formed as a result of members' sharing (visiting and staying at) each other's place and also participating in local events. In addition, pre-existing social networks and friendship ties that are usually place-specific are shown no longer underpinning individuals' participation in the new form of communities, which is instead increasingly driven by the associations' capacity to build new relationships (Parigi et al. 2013).

Inherent to the non-place-specific sharing practice and crucial to the community formation is that mutual trust between people is not established any more based on spatial proximities and face-to-face interactions. Instead, interpersonal trust is enabled and mediated by online information, often in the form of one's online profile and the rating and reputation created from crowdsourced information (Schor and Fitzmaurice 2015). As a result, what is not easy to be achieved offline before—creating trust and reducing uncertainties between a large group of strangers—is now normalized online, not only in a greater volume and intensity but also often prior to any actual interactions. However, while the necessity of face-to-face interactions for community building in the sharing economy remains debatable, the technology-mediated trust nonetheless has double-edged effects on the formation of new social relationships. The availability of information makes people become more and more calculating and selective about the unknown others with whom they would like to interact and share. As a result, "it made relationships easier to establish initially but it also weakened them after a certain threshold" (Nadeem et al. 2015: 19). In addition, significant problems with the reliability of crowdsourced information, such as free mutual vouching (Lauterbach et al. 2009), misinformation to avoid retribution (Overgoor et al. 2012), colluding and resultant inflated reputation (Wang and Nakao 2010), have also been identified that can easily undermine trustworthiness between people, reinforce skepticism, and consequently hinder community formation. All these ineluctably raise the concerns about the likelihood of forming new communities in the sharing economy.

Besides, another likely consequence of the technology-mediated trust is exacerbated discrimination (Schor 2016). This is because building trustworthiness online may foreclose the possibility of participation of those who cannot provide relevant and satisfactory information (Richardson 2015), hence constraining the sharing within a certain social or economic scope. This means that to a certain extent, the likelihood of individuals' participating in, and the extent to which they can benefit from, the sharing economy has already been constrained. This also means that it would not be easy and straightforward for one to build his or her trustworthiness through directly engaging in a sharing practice. Furthermore, the "fear of contagion" (Belk 2010: 726) that often hinders sharing with strangers may not diminish as a result of the ubiquitous sharing economy. Rather, because of the transparency of individuals' social status and ethnic information online, it may exert even stronger influences on one's decision, as it becomes much easier to choose the strangers with whom he or she would like to interact and share with. Such discrimination is not uncommon in the two most popular sectors of the sharing economy, namely car sharing (Sarriera et al. 2017) and accommodation sharing (Edelman et al. 2017), where prejudices toward clients regarding social class and race are evident.

2.3.3 Digital Platforms

We discussed earlier in this chapter that across different framings of the sharing economy, digital platforms have been repeatedly highlighted as the underpinning of the access to a great variety of resources and the sharing with extensively distributed strangers. In this sense, they can be objectively seen as a means to more efficient information dissemination and convenient connections, rather than exerting direct influences on the sharing practices. From this point of view, the role that the digital platform plays in the sharing economy is thus at most catalytic. However, the indirect impact of the unprecedented scale of both information and connections brought about by the digital platforms should not be taken for granted.

First, while the pervasive peer-to-peer connections enabled by the digital platforms are believed to be paving a way to a more decentralized and equitable economy (Botsman and Rogers 2010), it may at the same time risk depriving many existing, and even constraining new, opportunities for sharing. This is mainly because the dramatic decrease of transaction cost (Benkler 2004) and the increasingly growing market, of which both are created by the digital platforms, can lead to the 'crowding-out' phenomenon, where people's extrinsic motivation of economic gains starts overshadowing their intrinsic motivation of altruistic sharing (Hamari et al. 2016). In this case, not only can individuals be greatly motivated to monetize their underutilized assets, but also that what were generally not considered as commodities before are likely to be commodified. From the point of view of rational *homo economicus*, it is perhaps not unreasonable to maximize exchange values and minimize opportunity costs. Nevertheless, a higher risk of the 'crowding-out' phenomenon should not be neglected that over time it may change and shape individuals' mindset toward capitalizing on resources that are supposed to be, or have always been, shared within families, friends, communities, and beyond. Individuals' willingness to care and help others would otherwise likely to be weakened, and their altruistic and non-reciprocal motives of sharing suppressed.

While at the individual level the digital platforms may undermine one's intent of sharing, collectively they may lead to exacerbated inequality, which is the exact opposite of the claimed goal of the sharing economy. Specifically, the continuously growing digital connection network can contribute to amplifying the 'Piketty-effect' of the sharing economy (Frenken 2017; Piketty 2015; Frenken and Schor 2017). That is, a small group of well-off people who own a multitude of scarce consumer goods may become even wealthier, as they can more easily leverage a broader range of 'consumers' to generate more additional income. Not limited to this, the better educated, who are also often well-off, are found capturing, through the digital market, the low skilled job opportunities that were usually taken by the lower-educated workers in the past (Schor 2016). As a result, the digital platforms as a catalyst further skew the already unequally distributed resources and wealth toward the haves.

Second, the rapid advancement of digital technologies may continuously challenge the necessities of social connections stemming from the sharing practice. When access can be granted using one-time pin codes, coordination and distribution of the shared resources can be automated, and trust between individuals digitally mediated, the already limited opportunities for face-to-face interactions are likely to be rendered even more unnecessary. Taking for example the case of shared accommodation where social interactions are at least more likely to occur, a user of Airbnb may not have to collect keys from and return to the host in person. Instead, he or she can access the house and room freely by a temporary pin code during the stay. They may even not have to engage in any forms of online interactions beforehand, because the booking and other transactions can be conveniently arranged and efficiently settled using a deliberately designed digital platform. It is not difficult and unreasonable to imagine that similar cases may also apply to carpooling. This is not to mention that many business-to-consumer services, such as Mobike and Zipcar, are either initially designed to maximize convenience and efficiency with digital technologies at the expense of social interactions, or gradually phase out face-to-face interaction that was conceived as a key dimension at the beginning. Although we discussed before that not all sharing economy activities necessitate social connections and that not all participants desire interactions, the advancement of digital technologies nonetheless is likely to create new uncertainties for many socially oriented sharing practices.

However, it is noteworthy that at the same time the digital platforms have given rise to numerous virtual communities that are formed as a result of many online sharing activities, such as writing blogs, uploading photos and videos, and contributing to Wikipedia. In these communities, although individuals technically are not possessive of what is being shared, all members' generous contribution to, and joint enjoyment of, the virtual resources do raise a feeling of joint possession and subsequently cultivate a reverence for the community, which may lead to obligations of continuous sharing (Belk 2010, 2014a). Nevertheless, while these online communities seem to manifest what sharing entails as discussed at the beginning of this chapter, it remains an open question to what extent they can actually generate new social interactions and produce new social relations. With the advancement of digital technologies, they can be expected to continue to better foster online sharing based on indirect reciprocity, but probably not much more. Wittel (2011) also argues that while online sharing has the potential to create social interactions, it does not guarantee to trigger any social responses.

2.4 A Complex Picture

In the above analysis, we devote our focus to three key aspects of the sharing economy, namely, access, sharing with strangers, and digital platforms. We can see that even by singling out specific dimensions, the picture of sharing of each is neither simple nor uncomplicated. At least one paradox is suggested for each aspect. First, access can contribute to bringing together a variety of people because of both its economic benefits and symbolic values for identity building, but it may also constrain the formation of social bonds and communities for the very same reasons. Second, sharing with strangers potentially offers a vast opportunity for extensive social connections and interactions, but the internet-based and digitally mediated trust between individuals can hurdle the production of new social relations and even exacerbate discrimination and segregation. Third, digital platforms as the underpinning of the sharing economy enable convenient and efficient connections and are supposed to greatly facilitate sharing, but the sheer scale of the connections coupled with the advancement of digital technologies can suppress individuals' altruistic motives and diminish possibilities of social connections. On their own, these three aspects underscore almost every single sharing economy platforms, and they are often interconnected. Placed together with their paradoxes, the fuller picture of sharing in the sharing economy has become even more complex.

In a couple of places, we discuss that different motivations ranging from the purely altruistic to the completely egoistic always coexist and affect people's engagement in the sharing economy. At a finer-grained level, what plays a crucial role in forming individuals' positive attitudes is often different from what directly drives their actual participation (Hamari et al. 2016). And with respect to different sectors of the sharing economy, Böcker and Meelen (2017) also showed that the combination of motivations for participation can vary considerably. And they can also contrast between different social-demographic groups and between users and providers. Furthermore, in the above analysis we argue that the combination of motivations can be influenced by the changes within the larger environment. For instance, one's altruistic intent of sharing can be overshadowed by others' pursuit of economic incentives, which are likely to be greatly amplified by the vast emerging digital market. Or conversely, one may become more prosocial because of others' generous and continuous sharing, even if his and her initial intent is primarily a pursuit of economic incentives. Considering these nuances, it is probably reasonable to conclude that in the sharing economy, the altruistic sharing is more than often mixed and intertwined with what Belk (2014a) refers to as 'pseudo sharing'.

We also show in the analysis that the social effects of the sharing economy are equally not straightforward, especially intertwined with economic incentives. On the one hand, the social and economic gains usually conflict with each other. The potential of many sharing platforms to foster altruistic sharing and generate communal benefits can be conditioned and even substantially shrunken by the market mediation. And in some cases, even if interactions between individuals are desirable and not driven by the pursuit of economic incentives, they neither necessarily produce new relations, nor do they create new bonds and trust. On the other hand, social benefits can also complement with, and may even be predicated on, the economic ones. The coworking space, for example, is first a for-profit platform that extracts rents of desks or office rooms from workers, but on this basis, it rewards them with opportunities for socialization and collaboration that are engendered by the physical proximity of a large group of diverse workers. From this perspective, one could argue that it is often the alignment of economic benefits with the social goals that fosters sharing and sustains many sharing economy platforms.

The social effects of the sharing economy are also entangled with its environmental influences. For some sharing economy platforms, such as Airbnb and Couchsurf, of which social gains are claimed among the key objectives (Schor and Attwood-Charles 2017), the pursuit of the social goals itself, regardless of the actual achievement, by itself entails additional burdens on the environment—it first necessitates traveling and may further incentivizes more people to travel more frequently. For other platforms such as car sharing, on the contrary, while its social benefits remain questionable (Bardhi and Eckhardt 2012), the direct environmental gains are almost indisputably positive: considerable reduction of carbon emission was reported (Chen et al. 2016). Also noteworthy are the 'ripple effects' (Schor 2016) of the sharing economy, when a new activity may set into motion a series of other changes and influences in the system. For instance, economic gains from capitalizing on idle capacities may collectively contribute to shift of income and more uneven distribution of welfare across social classes (Frenken and Schor 2017). This can lead not only to a further weakening of existing social connection network, but also to the further increase in the beneficiaries' consumption level, which in turn may undermine environmental sustainability that has always been presumed by the sharing economy.

2.5 A Need for Enhanced Sharing

Our arguments in this chapter demonstrate that there is a need to design sharing activities, practices, and systems to further amplify their social benefits. Sharing in the real world, as our arguments demonstrate, is an impure exercise laced with many different motivations—some altruistic, others characterized by indifference, and others perhaps, egoistical. To persist with this status quo is therefore no different than either to acquiesce to the market of the sharing economy, or to accept the random happenstance of what dominant sharing practices produce. However, to consider sharing as a specific target of design may serve to conserve, or perhaps even, accentuate, many of its prosocial traits. In turn, these qualities can lead to other socially constructive agencies and practices.

References

F. Bardhi, G.M. Eckhardt, Access-based consumption: the case of car sharing. J. Consum. Res. **39**(4), 881–898 (2012). https://doi.org/10.1086/666376

R. Basselier, G. Langenus, L. Walravens, The rise of the sharing economy. Econ. Rev. (iii), 57–78 (2018)

P. Beck, M. Hardie, N. Jones, A. Loakes, The feasibility of measuring the sharing economy: November 2017 progress update. Accessed Office of National Statistics website (2017), https://www.ons.gov.uk/economy/economicoutputandproductivity/output/articles/thefeasibilityofmeasuringthesharingeconomy/november2017progressupdate

R. Belk, Possessions and the extended self. J. Consum. Res. **15**(2), 139–168 (1988). https://doi.org/10.1086/209154

R. Belk, Why not share rather than own? Ann. Am. Acad. Polit. Soc. Sci. **611**(1), 126–140 (2007). https://doi.org/10.1177/0002716206298483

R. Belk, Sharing. J. Consum. Res. **36**(5), 715–734 (2010). https://doi.org/10.1086/612649
R. Belk, Sharing versus pseudo-sharing in Web 2.0. Anthropol. **18**(1), 7–23 (2014a), https://doi.org/10.1080/09720073.2014.11891518
R. Belk, You are what you can access: sharing and collaborative consumption online. J. Bus. Res. **67**(8), 1595–1600 (2014b). https://doi.org/10.1016/j.jbusres.2013.10.001
Y. Benkler, Sharing nicely: on shareable goods and the emergence of sharing as a modality of economic production. Yale Law J. **114**, 273 (2004)
L. Böcker, T. Meelen, Sharing for people, planet or profit? analysing motivations for intended sharing economy participation. Environ. Innov. Soc. Trans. **23**, 28–39 (2017). https://doi.org/10.1016/j.eist.2016.09.004
C. Boehm, The natural selection of altruistic traits. Hum. Nat. **10**(3), 205–252 (1999)
R. Botsman, The sharing economy lacks a shared definition (2013), https://www.fastcompany.com/3022028/the-sharing-economy-lacks-a-shared-definition
R. Botsman, R. Rogers, *What's Mine is Yours: the Rise of Collaborative Consumption* (Harper Collins, New York, 2010)
T.D. Chen, K.M. Kockelman, J.P. Hanna, Operations of a shared, autonomous, electric vehicle fleet: implications of vehicle & charging infrastructure decisions. Transp. Res. Part A: Policy Pract. **94**, 243–254 (2016). https://doi.org/10.1016/j.tra.2016.08.020
S.R. Davies, in *Hackerspaces: making the Maker Movement* (Polity, Malden, MA, 2017)
G.M. Eckhardt, F. Bardhi, The sharing economy isn't about sharing at all. in *Harvard Business Review* (2015). https://hbr.org/2015/01/the-sharing-economy-isnt-about-sharing-at-all
B. Edelman, M. Luca, D. Svirsky, Racial discrimination in the sharing economy: evidence from a field experiment. Am. Econ. J.-Appl. Econ. **9**(2), 1–22 (2017). https://doi.org/10.1257/app.20160213
M. Felson, J.L. Spaeth, Community structure and collaborative consumption: a routine activity approach. Am. Behav. Sci. **21**(4), 614–624 (1978). https://doi.org/10.1177/000276427802100411
A.P. Fiske, *Structures of Social Life: the Four Elementary Forms of Human Relations: communal Sharing, Authority Ranking, Equality Matching, Market Pricing* (Free Press, New York, 1991)
K. Frenken, Political economies and environmental futures for the sharing economy. Philos. Trans. R. Soc. A: Math. Phys. Eng. Sci. **375** (2017). https://doi.org/10.1098/rsta.2016.0367
K. Frenken, J. Schor, Putting the sharing economy into perspective. Environ. Innov. Soc. Trans. **23**, 3–10 (2017). https://doi.org/10.1016/j.eist.2017.01.003
L. Gansky, *The Mesh: why the Future of Business is Sharing* (Penguin, New York, 2010)
J. Hamari, M. Sjoklint, A. Ukkonen, The sharing economy: why people participate in collaborative consumption. J. Assoc. Inform. Sci. Technol. **67**(9), 2047–2059 (2016). https://doi.org/10.1002/asi.23552
T. Ikkala, A. Lampinen, in *Monetizing Network Hospitality: hospitality and Sociability in the Context of Airbnb* (2015)
H. Jarvis, Sharing, togetherness and intentional degrowth. Prog. Hum. Geogr. **43**(2), 256–275 (2019). https://doi.org/10.1177/0309132517746519
N.A. John, Sharing and Web 2.0: the emergence of a keyword. New Media Soc. **15**(2), 167–182 (2013a). https://doi.org/10.1177/1461444812450684
N.A. John, The social logics of sharing. Commun. Rev. **16**(3), 113–131 (2013). https://doi.org/10.1080/10714421.2013.807119
N.A. John, *The Age of Sharing* (Polity Press, Meldon, MA, 2017)
D. Lauterbach, H. Truong, T. Shah, L. Adamic, *Surfing a Web of Trust: reputation and Reciprocity on Couchsurfing. com.* Paper presented at the 2009 International Conference on Computational Science and Engineering (2009)
S.J. Lawson, M.R. Gleim, R. Perren, J. Hwang, Freedom from ownership: an exploration of access-based consumption. J. Bus. Res. **69**(8), 2615–2623 (2016). https://doi.org/10.1016/j.jbusres.2016.04.021

Y. Lin, Fridges stocked with food in Tampines block to help boost kampung spirit. in *The Straits Times* (2017). https://www.straitstimes.com/singapore/fridges-stocked-with-food-in-tampines-block-to-help-boost-kampung-spirit

C.J. Martin, The sharing economy: a pathway to sustainability or a nightmarish form of neoliberal capitalism? Ecol. Econ. **121**, 149–159 (2016). https://doi.org/10.1016/j.ecolecon.2015.11.027

S. Nadeem, J.B. Schor, E.T. Walker, C.W. Lee, P. Parigi, K. Cook, On the sharing economy. Contexts **14**, 12–19 (2015)

J. Overgoor, E. Wulczyn, C. Potts, *Trust Propagation with Mixed-Effects Models.* Paper presented at the Sixth International AAAI Conference on Weblogs and Social Media (2012)

J. Owyang, C. Tran, C. Silva, *A Market Definition Report—The Collaborative Economy* (2013), http://www.collaboriamo.org/media/2014/04/collabecon-draft16-130531132802-phpapp02-2.pdf

L.K. Ozanne, P.W. Ballantine, Sharing as a form of anti-consumption? an examination of toy library users. J. Consum. Behav. **9**(6), 485–498 (2010). https://doi.org/10.1002/cb.334

L.K. Ozanne, J.L. Ozanne, A child's right to play: the social construction of civic virtues in toy libraries. J. Pub. Policy Mark. **30**(2), 264–278 (2011). https://doi.org/10.1509/jppm.30.2.264

I. Pais, G. Provasi, Sharing Economy: A Step towards the Re-Embeddedness of the Economy?, in *Stato e mercato, Rivista quadrimestrale*, pp. 347–378 (2015), https://doi.org/10.1425/81604

P. Parigi, B. State, *Disenchanting the World: the Impact of Technology on Relationships.* Paper presented at the Social Informatics: 6th International Conference, SocInfo 2014 (Barcelona, Spain, 2014)

P. Parigi, B. State, D. Dakhlallah, R. Corten, K. Cook, A community of strangers: the dis-embedding of social ties. PLoS ONE **8**(7), e67388 (2013). https://doi.org/10.1371/journal.pone.0067388

R. Perren, L. Grauerholz, Collaborative consumption. Int. Encycl. Soc. Behav. Sci. **4**, 139–144 (2015). https://doi.org/10.1016/B978-0-08-097086-8.64143-0

T. Piketty, About capital in the twenty-first century. Am. Econ. Rev. **105**(5), 48–53 (2015). https://doi.org/10.1257/aer.p20151060

J.A. Price, Sharing: the integration of intimate economies. Anthropologica, 3–27 (1975), https://doi.org/10.2307/25604933

L. Richardson, Performing the sharing economy. Geoforum **67**, 121–129 (2015). https://doi.org/10.1016/j.geoforum.2015.11.004

J.M. Sarriera, G.E. Álvarez, K. Blynn, A. Alesbury, T. Scully, J. Zhao, To share or not to share: investigating the social aspects of dynamic ridesharing. Transp. Res. Rec. **2605**(1), 109–117 (2017). https://doi.org/10.3141/2605-11

T. Schaefers, S.J. Lawson, M. Kukar-Kinney, How the burdens of ownership promote consumer usage of access-based services. Marketing Letters **27**(3), 569–577 (2016). https://doi.org/10.1007/s11002-015-9366-x

J. Schor, Debating the Sharing Economy, 1–13. Accessed Great Transition Initiative website (2016), http://greattransition.org/publication/debating-the-sharing-economy

J. Schor, W. Attwood-Charles, The "sharing" economy: labor, inequality, and social connection on for-profit platforms. Sociol. Compass **11**(8), e12493 (2017)

J. Schor, C. Fitzmaurice, Collaborating and connecting: the emergence of the sharing economy, in *Handbook of Research on Sustainable Consumption*, ed. by L.A. Reisch, J. Thøgersen (Edward Elgar Publishing, UK, 2015)

D. Selloni, New forms of economies: sharing economy, collaborative consumption, peer-to-peer economy. in *CoDesign for Public-Interest Services* (Springer, 2017), pp. 15–26

C.B. Stack, *All Our Kin: strategies for Survival in a Black Community* (Harper & Row, New York, 1974)

C.A. Steinkuehler, D. Williams, Where everybody knows your (screen) name: online games as "third places". J. Comput.-Mediat. Commun. **11**(4), 885–909 (2006). https://doi.org/10.1111/j.1083-6101.2006.00300.x

K. Stokes, E. Clarence, L. Anderson, A. Rinne, *Making Sense of the UK Collaborative Economy* (Nesta, London, 2014)

E. Sung, H. Kim, D. Lee, Why do people consume and provide sharing economy accommodation?—a sustainability perspective. Sustainability **10**(6), 2072 (2018). https://doi.org/10.3390/su10062072

D. Thomas, Intra-household resource allocation: an inferential approach. J. Hum. Resour. **25**(4), 635–664 (1990). https://doi.org/10.2307/145670

I.P. Tussyadiah, Factors of satisfaction and intention to use peer-to-peer accommodation. Int. J. Hosp. Manag. **55**, 70–80 (2016). https://doi.org/10.1016/j.ijhm.2016.03.005

Y. Wang, A. Nakao, Poisonedwater: an improved approach for accurate reputation ranking in P2P networks. Future Gener. Comput. Syst. **26**(8), 1317–1326 (2010). https://doi.org/10.1016/j.future.2009.05.001

T. Widlok, Sharing by default? outline of an anthropology of virtue. Anthropol. Theory **4**(1), 53–70 (2004). https://doi.org/10.1177/1463499604040847

M. Williams, J. Hall, Hackerspaces: a case study in the creation and management of a common pool resource. J. Inst. Econ. **11**(4), 769–781 (2015)

A. Wittel, Qualities of sharing and their transformations in the digital age. Int. Rev. Inf. Ethics **15**(9), 3–8 (2011)

Chapter 3
Sharing by Design

Sometime early in the writing of this book, we found ourselves meeting in a café to discuss our progress. As usual, Ye brought his computer and I, my handwritten notes. We sat down, facing each other, with our beverages, laptop, notes, and notepad, all on a small table between us (see Fig. 3.1a). At first, our discussion went well enough; we discussed the outlines of our own progress, the impending schedule for writing, and other matters that overlap with this book project. But as the discussion progressed, we found it necessary to begin referring to what we had brought for this meeting. Ye would need to rotate his laptop in my direction whenever he wanted to share his ideas with me; but if I could view the screen, this meant that he could not. On my side, I would share my handwritten notes with him; but this meant that if he could read my notes, then I would be trying to read upside down. And there and then, there was the small but constant challenge of adjusting our research materials—laptop, notes, notepad, and pens—on the small rectangular table whenever one side tried to communicate his ideas to the other side. No one really wanted to spill coffee and tea on anything.

This went on for a while until we realized the irony of our predicament. Here were two authors writing on sharing but having a great difficulty in sharing their ideas! One of us then shifted his chair so that we no longer sat face-to-face, but instead, perpendicularly to each other (see Fig. 3.1b). This new configuration changed the default layout of the café but allowed us to share more effectively: I could read his laptop screen—so could Ye—and Ye could view my notes—as could I. And instead of constantly re-adjusting the research materials on the table (and the hot beverages), we needed only to activate one corner of the table while leaving the rest for other belongings. By simply changing the original configuration, we had made our sharing more effective and efficient, and we might add, kept our research materials from the harm that toppled beverages could inflict.

This little episode shows that sharing knowledge is intuitive. But it is less intuitive to improve sharing by design. Clearly, not all sharing configurations are so easily improved by moving a chair around, or else re-positioning ourselves and our

J. K. H. Chan and Y. Zhang, *Sharing by Design*, SpringerBriefs in Applied Sciences and Technology, https://doi.org/10.1007/978-3-030-43569-1_3

a b

Fig. 3.1 Café (*Source* Authors)

belongings differently. Yet all sharing configurations are first improved by an initial recognition for more robust sharing. This recognition is essential whether one is trying to communicate an ideal of what sharing can attain, or whether one is trying to visualize a sharing culture. This recognition constitutes the beginning of sharing by design.

In this chapter, we will systematically review the present discourse on designing sharing systems, emphasizing on the notable paradigm of 'design by capital' and 'design by commoning'.

3.1 Why Is Sharing Ascendant Today?

The present study is guided by a need to design new sharing systems in cities today. But why there is a need for these systems? Why, especially in cities, are new sharing systems urgently needed? One explanation points to the limitations of neoliberal, growth-dependent paradigm of urban planning (Rydin 2013). This paradigm drives the relentless growth of cities, which both stabilizes and advances capitalism (Harvey 2010). However, much of this growth has been attained at the expense of severely degrading the natural environment (Standing 2019), while at the same time, positions investors and developers—rather than citizens—at the helm of shaping the neoliberal city (Mayer 2010: 36). As a result, neoliberal cities are characterized internally by private developments and their imposed enclosures (Hodkinson 2012), and externally on the frontiers of an ever-expansive urban process, by urbanization that threatens land-grabs, eviction and displacement, and ultimately the expulsion of people from their homes and indigenous lands (Sassen 2014). This has resulted in nothing less than a deepening urban crisis across the world, which has also accreted into an inhospitable planetary urbanization: that in spite of an ever-expanding program of city-building all across the world, an ever larger segment of humanity remains alienated from the city. The struggle for affordable housing, for instance, has become a key flashpoint of urban discontent across the world.

In Badiou's (2016) estimation, there are presently more than two billion people—and rising—who are left out of the global political economy today. These people are neither workers nor consumers, and who are outside the marketplace of jobs, goods, and services, and counted for nothing by the market society. Many of them constitute the multitude that has been 'expelled' from their dispossessed homes and lands (Sassen 2014), and subsequently arrive in cities in search of shelter and work (Saunders 2010). Compounding this reality, the acceleration of automation, work obsolescence by artificial intelligence, and market turbulences in recent decades have drastically unsettled an otherwise relatively stable trend of work and secured employment in advanced capitalist economies. An entire social class of the 'precariats' has arisen in many cities in the so-called Global North today (Standing 2011). To make matters worse, this trend has coincided with a spade of successive state cutbacks for many public welfare programs and amenities, resulting in a deprived urban population in many places with little recourse except for market solutions that it cannot afford. For a growing group in cities then, access to urban resources such as affordable housing, education, mobility, healthcare, childcare, parks and more recently, nutritious food and clean water, has become increasingly precarious and scarce.

This context then begins to explain why sharing has become ascendant. Initially, the sharing economy had projected a promising form of collaborative consumption that exhorts affordable access rather than ownership of expensive assets, which could have mitigated the worst of the urban crisis. However, and in spite of the lofty reference to sharing, corporations driving the sharing economy have been criticized for focusing on profit maximization, and evading regulation while equity and sustainability impacts of sharing are neglected (Frenken 2017: 2). Although researchers have noted the presence of social and environmental motivations—such as altruism and environmentalism—in participants of the sharing economy (Böcker and Meelen 2017), nevertheless taken as a whole, the sharing economy, for the various reasons delineated (see Chap. 2), cannot be counted on to mitigate this urban crisis. A different socio-political formation to re-organize production and consumption, and to pool and manage critical urban resources, has become necessary either to supplement the sharing economy, or else to overtake it completely in due course.

But this new formation has to be designed. Efforts to materialize this new socio-political formation can be categorized as a form of *transition design*, which refers to design-led efforts for societal transition toward a more sustainable future (Irwin 2015). Transition designers postulate that design can contribute meaningful solutions to complex and wicked problems (Irwin 2015). Unlike market-oriented and short-term design outcomes, transition design approaches are oriented to longer time horizons, as they are also informed by visions of a sustainable future (Escobar 2017). Transition design focuses on making substantive and corrective changes to everyday practices that undermine this future, and moreover, relies on a place-based approach and actual sites for design intervention (Irwin 2015). Seen in this light, sharing by design is proposed as one specific strategy within the broad framework of transition design.

3.2 Reflexivity in Design: Ought Sharing Be Designed?

Even so, to remain accountable, design requires reflexivity—that is, a critical distance to permit the responsible appraisal of design actions and their consequences. This reflexivity has become paramount because we live in a design culture today (Krippendorff 2006), where much of contemporary life is either spent designing, or else mediated by artificial artifacts, processes, and systems. A large part of human activities now revolves around the use of technology to transform possibilities into realities (Krippendorff 2006: 72). In this way, design is fast becoming a primary pathway for self-actualization. In this design culture catalyzed by powerful technologies, it is neither simple nor easy to distinguish between what to design and what not to design. As Horst Rittel once suggested, to design is often to invoke "the curse of feasibility" (Protzen and Harris 2010: 223), where 'I can' implies 'I ought'. One then wonders about the reflexive capacity that must exist to rein in the growing power to design everything.

And so even if transition design may be deemed necessary, a careful inquiry on sharing by design still has to begin with this question: 'ought sharing be designed?'. Here, 'ought' connotes two different meanings (Hunter and Nedelisky 2018: 141–142). First, 'ought' connotes the ethical and the prescriptive, where, for instance, one says that sharing ought to be designed in order to attain certain desirable or ideal goals. Second, 'ought' connotes the practical or the prudential, where, for instance, one remarks that if resources are to be distributed, then one ought to design a sharing system to do just that. This second connotation does not invoke the ethical but the practical. Here, we are interested in both meanings of 'ought'.

In relation to the ethical connotation of 'ought', we suggest that to design sharing can mean designing structures and configurations that often threaten, or else subordinate, the very volitional agency deemed ethically valuable in sharing relations. What becomes of sharing if one has been impelled to share? We will discuss the moral threats to sharing in greater details in Chap. 5. But at least for now, it is clear that these threats are not only morally problematic, but they can also undermine the ethos of sharing in the long run. However, in relation to the practical connotation of 'ought', we suggest that to design sharing is often tantamount to institutionalizing sharing, which is an attempt to systematize sharing behaviors and relations—rendering them stable, repetitive, and predictable within a structured configuration (see Goodin 1996). But it is uncertain whether sharing—which is in part predicated on volitional agency and free associations between individuals—can even be sustained in such a structured and institutionalized setting. It does appear that for sharing configuration to be sustained, it has to at least avoid the twin pitfalls of threatening the volitional agency to share, and the undesirable effects of institutionalized sharing.

In sum, sharing by design has to attend to three fronts. First, to have the capacity or capability to design for sharing is not, by default, a normative decision to design it: 'Is' does not imply 'ought'. Instead, the normative decision to design sharing has to be made in reference, at the very least, to a morally worthy goal. Second, attempts of sharing by design have to consider the impacts of design on the moral quality and development of volitional sharing agencies, relations, and ecologies. These two fronts suggest that although frequently neglected, ethics nevertheless plays a pivotal role in sharing by design. And finally, these attempts have to consider the unintended impacts of structures and configurations that are designed to motivate, regulate, or sustain sharing agencies and practices. While they may regulate and systematize sharing behaviors and practices, these structures and configurations can nevertheless limit, if not also undermine, a further propensity to share. This last front suggests the necessity to consider sharing by design through the approach of systems thinking, where the design has to be examined in relation to the broader context where it would be emplaced. The systems approach for sharing by design will be further explored in Chap. 4.

There are presently two salient paradigms of sharing by design. We refer to the first paradigm as 'design by capital', and the second paradigm as 'design by commoning'. We elaborate on each of these paradigms.

3.3 Design by Capital

'Design by capital' is a design paradigm that aims to systematize sharing behaviors for business-oriented goals in the sharing economy. Driven by commercial interests, design is used to materialize sharing configurations that can enable or facilitate sharing behaviors and activities for the goal of yielding financial returns. To do this, design by capital usually relies on professional researchers and designers to first study the existing sharing practices and then to propose innovative improvements to sharing configurations. For example, to enhance responsible sharing behaviors, new technologies of geo-tagging have been proposed and then implemented on shared bikes and scooters. In turn, this reduces uncertainty in business operations, which can better ensure profits. Another example is the wholesale transformation of old commercial spaces into a new co-working space to tap on the rising market demand for flexible workspaces. Central to these design processes is the need to discover and specify problems encountered by the sharing practices, and subsequently to scan for untapped market opportunities or latent potentials for which design solutions could be found to exploit them.

As discussed, the co-working space is one salient example of design by capital. In light of the emerging nomadic and independent work culture across the world, a vast market has opened up for temporal and flexible workspaces. However, a design problem for many co-working configurations is how to exploit the potentials of co-location and co-presence in order to motivate the formation of collaborative relations between workers from different backgrounds (Merkel 2017). Spatial design solutions that can trigger the preconditions of collaboration—for instance, rendering a frequently used staircase or pantry more conducive for unplanned or serendipitous interactions (see Wagner and Watch 2017)—then become important. Importantly, designers have started to acknowledge that physical spaces have to be specifically designed for systematic sharing activities to take place, and that there are also resources that can be more effectively shared in real and material arenas (Piccinno 2018). Even so, systematic design knowledge that can 'nudge' workers to interact more actively, or to catalyze the formation of organic solidarity between workers, remains underdeveloped.

Our focus here is not on co-working, which has already been examined on numerous fronts (Babb et al. 2018; Gandini 2015; Mariotti et al. 2017; Merkel 2017; Olma 2016). Instead, we focus on the emerging phenomenon of co-living spaces, which remains relatively underdeveloped. This category of shared accommodation between strangers is set for further growth because housing has become increasingly scarce and unaffordable for workers in many cities (Bhatia and Steinmuller 2018). Furthermore, co-living, as the provision of shelter, is also a sharing configuration deemed to be more demand inelastic relative to co-working. To further illustrate our arguments on design by capital, we will examine the case study of the Starcity San Jose co-living facility.

3.4 A Concise Case Study: Co-living in Starcity San Jose

Co-living is a rapidly growing sector within the sharing economy (Knight Frank 2019). As an architectural and real-estate asset, co-living is a form of shared housing that combines private living spaces with shared communal facilities (Knight Frank n.d.). This configuration aims to promote social engagement between residents and foster the development of a community, which is also the main unique feature of co-living relative to rental housing and service apartments (Knight Frank 2019). Following this, there are future plans to integrate both co-living and co-working in one mutually reinforcing spatial ecosystem, which insofar, as real-estate developers are concerned, is an effective way to expand, as well as retain, member base and therefore, also revenue (Knight Frank 2019).

The rapid growth of co-living today is driven by many factors, but two primary factors stand out. First, work has become more nomadic and project-based, and independent workers now move between cities for work. As discussed, this has triggered the need for the co-working space. But in tandem, this has also led to a similar demand for the co-living space. Importantly, the co-living configuration has been specifically designed to cater to the flexible and temporal occupancy patterns desired by independent workers. This arrangement also includes a convenient suite of services (e.g. janitorial services, utilities, and internet access) that is integrated within a monthly rent. Moreover, many of these independent workers are new to their arrival city, and the co-living configuration then becomes a significant shared venue where they can meet and connect with new people on a regular basis (Knight Frank 2019).

Second, the economic profile of this new generation of workers is diametrically distinct from their predecessors. It is unclear if the workers' preference for access, rather than the ownership of assets, stems from the nature of their nomadic and short-term work, or the rising unaffordability of housing in relation to their stagnant wages. Nevertheless, it is clear that these workers constitute an increasingly important aspect of what Lazzarato (2011) refers to as the debt economy—a political economy sustained by low growth, stagnant wages but high debts, and where politics is driven by exploiting the power differentials between creditors and debtors. For these workers, loans have become the primary means to meet the major needs (Hardt and Negri 2012: 10), and debt repayment on these loans then consumes a large part of their stagnant wages. In evading this foreseeable debt bondage, it is no small wonder that workers today prefer as much flexibility and freedom from debt as possible. In sum, given this emerging nature of their work and economic profile, co-living then becomes an appealing option—not so much the ideal form of urban accommodation but one that is at least practical and tailored to the workers' needs.

At the time of writing, Starcity San Jose has been deemed as the largest co-living building in the world (Holder 2019). It has also been approved for construction by the city. Our approach here relies on the constructive evaluation of architectural design through the perspective of real-estate agency and rationality (see Martin 2015). By reviewing the dynamics of how real-estate agencies could develop into a work of architecture for co-living, this case study aims to reveal, even if limitedly, how a specific design by capital is produced. And by evaluating this project from the calculations of real-estate rationality, we demonstrate that contrary to common intuition, the logic of sharing accommodation is not incompatible to the logic of a profit-driven real-estate operation. In other words, sharing is an efficient use of scarce urban land, but this efficiency also enhances real-estate profits. Design by capital for co-living then anticipates the prospect of profits in an under-supplied market segment, further reinforced by the real-estate efficiency brought about by sharing. This preliminary finding then projects that design by capital is likely to constitute a key approach for driving the design of shared accommodation.

To begin, in May 2019, an addendum titled, 'Starcity Residential Project [File No. SPA17-023-01]' was filed in addition to an approved development project titled, 'Bassett Street Residential Project (Aviato)' [File No. Sp17-023 and T17-026], in the city of San Jose, California, USA (San Jose City Hall 2019). In this addendum, the developer Starcity proposed turning the original private housing tower into a co-living development. Originally developed as an 18-story tower with 302 residential units on top of some 7,182 sq. ft of ground floor restaurants and retail, a fitness area, and pool-deck terrace on the second and the seventh floor, respectively, Starcity instead proposed to transform these 302 residential units into 803 co-living residential units with a nearly 50% reduction of the original floor area dedicated to restaurants and retail. In the addendum [File No. SPA17-023-01], the report states that 803 co-living units are, conservatively speaking, equivalent to 573 standard residential units (which, in other words, is also an additional increase of 271 standard residential units over the original proposal).

Conceived as a project that could offer greater accessibility to workers in the city, such co-living real-estate 'products' target middle-class workers in expensive cities on one hand, and on the other hand, aims to provide a community to individuals who feel disconnected or isolated within existing housing types (Dishotsky n.d.). Starcity's co-living buildings boast private bedrooms and bathrooms, with a centralized kitchen that is shared between 6 and 18 or 20 individuals (Dishotsky n.d.). These co-living units come fully furnished with enterprise grade Wi-Fi, and utilities and janitorial services in common areas are included in a convenient monthly bill. Importantly, 'members' can apply to reside in a co-living unit online, and can move in as quickly as within an hour—what Starcity's CEO Jon Dishotsky described as the seamlessness of, "from (web)site to sleep" (Dishotsky n.d.). And through a downloadable App, members can communicate with the management of the co-living building. Starcity also studies the demographics of its members, and discovered that people who live in co-living buildings vary: from seniors in their 70s to workers in their 30s. In sum, Starcity offers its members an offering not found anywhere else, which Dishotsky (n.d.) refers to as "Basics (shelter and amenities) plus Services (everything taken care of) plus Wow (unexpected treats and personalized treatment)".

While the exact reason for the modification toward co-living is unclear, it is however clear that the numbers of leasable or sellable residential unit equivalent have just been increased by nearly 90% through the transformation of the original residential units into shareable units. It does appear then, prima facie, that the reformulation to co-living is accompanied by a significant real-estate gain. Following this, it also appears that densification by sharing—that is, enabling more people to be packed within the same plot footprint through a co-living configuration—is not incompatible to the business logic of expanding profits through an intensification of urban land-use. In this way, the social and spatial factors that make sharing more likely then reinforce the logic that renders real-estate profits more probable. At least for the co-living category, profiteering can be reinforced by sharing.

While it is still too early to assess the effectiveness of design by capital for the co-living configuration, it is nevertheless possible to make two general observations. First, the co-living model represents a prototypical case of designing a complex sharing configuration systematically from the ground up. The overall design goal is to improve the feasibility of this sharing configuration, and this has been achieved by acknowledging a concerted interplay of different design strategies—namely but at least, architectural, social, managerial and financial—that are all aligned to this goal. Specifically, these design strategies have to mutually reinforce each other. For instance, the architectural strategy can promote flexible and adaptive shared spaces, which can encourage different forms of social interactions and events to take place that in turn require other social strategies to promote them. This overview of understanding how design strategies can mutually reinforce each other then presumes what Alexander (1964: 55) once referred to as a "selfconscious" design process—a process that aims to delineate and distinguish the different design variables and strategies at play, and to categorize them within a hierarchy composed of interacting relationships. To formulate this a little differently, the 'selfconscious' process aims to master the design process by demystifying it: to identify the key constituents of the design, to map how they interact, and then to further discover how to modify them in order to enhance the original goal.

Second, learning is key for any 'selfconscious' design process. Because the primary constituents and relations in any complex design system can now be explicated, it has also become possible to discuss and criticize them. Through the feedback generated by this discursive process, the designer can then learn how to improve on his or her design in subsequent iterations. This form of learning, when backed by the vast financial resources driving the co-living configuration and facilitated in real time by learning algorithms embedded in sharing platforms today, means that design learning can accelerate rapidly. If nothing else, design by capital reveals the unsettling prospect of designers learning enough about the mechanisms of sharing in co-living configurations to impel them.

3.5 Design by Commoning

In contradistinction, 'design by commoning' is a bottom-up, democratic, and participatory paradigm to design sharing configurations. In this paradigm, individuals design with, and alongside, each other. Professional designers do not design so much for these individuals, but instead, they design with these individuals (see Schwarz 2016: 132). In design by commoning, everyone who participates in the process of commoning becomes first *a commoner* and then the designer. Design by commoning therefore denotes a process where a commoner also designs. On this, Ostrom's (2006: 90) seminal work on common-pool resources represents a prototypical example of design by commoning, where members of a community come together to jointly design a collective institution that can regulate the behaviors of resource use and to manage an active commoning process.

But what exactly is commoning? And how is commoning different when it is practiced in the city—a context categorically different from the commons of meadows, fisheries, and woodlands found in the work of Ostrom (2006)? First and conceptually, Hardt and Negri (2012: 105) define commoning as a process that makes certain resources *common*. In turn, what is common is accessible and shared by all. Commoning is therefore an act of opening up resources that were once privatized, and transforming these resources into *a commons* that would be jointly owned, managed, developed, and sustained by the commoners. In tandem, commoning suggests the continuous making and remaking of the commons through a shared practice (Bresnihan 2016: 95)—a continuous process of winning back and expanding what should be held in common (Hardt and Negri 2009: ix).

Framed this way, commoning in the city is the process that creates new spatio-temporal and ethical formations concerned with ways of living together that resist privatization and the individualization of life (Dawney et al. 2016: 1). Standing (2019) argues that this form of commoning provides social amenities essential to normal living and contributes to an individual's social income by lowering the cost of attaining or maintaining an acceptable standard of living. Commoning in the city also produces new relations between individuals—strangers—who otherwise have little to do with each other (De Angelis 2016). It is, therefore, consistent to infer that unlike the natural resources that form the locus of Ostrom's (2006) commons, relations between commoners—and the services, amenities, and goods that these relations can collectively produce—constitute the kernel of commoning in the city.

On this, Kornberger and Borch (2015) suggest that *the urban commons* comprises resources that can increase in value even as they are consumed. One could imagine, for instance, how a cooperative childcare initiated by parents might begin with just the consumption of babysitting services but then expand to socializing and leisure activities organized by young families (Katrini 2018: 432). And yet again unlike the non-excludable but rivalrous consumption of natural commons, which tends to result in 'the tragedy of the commons' (Hardin 1968), the urban commons, however, bears a plausible prospect for 'the comedy of the commons' (Rose 1986)—where free and open access to a certain resource leads to a greater participation of its use, and in turn, increased usage creates greater social value (Frischmann 2012).

3.6 Co-housing: An Example of Design by Commoning

While there has been a shift from thinking of the commons as site or resources to commoning as a social practice (see Dawney et al. 2016: 4), much can also be said that the activity of commoning has to first presume a commons: "we need a commons to common" (Standing 2019: 347). Co-housing is therefore a relevant commons to further examine design by commoning.

What is co-housing? Among different definitions, one definition states that co-housing is an intentional community where residents actively participate in the design, planning, and governance of their living spaces (Jarvis 2019). Another definition suggests that co-housing is a kind of living arrangement that seeks to facilitate a high quality of life in ecologically sustainable homes and vibrantly interactive neighborhoods, and designed by following a cooperative ethos (Sargisson 2018). Common features of the co-housing types include a common house, where a large and shared kitchen is situated, and where mailboxes are also located (Sargisson 2018). One case study on co-housing emphasizes these features as venues for valuable micro-interactions between residents that can lead to greater social interactions (Chatterton 2016). And except for private spaces (e.g. bedrooms and bathrooms), most of all other spaces are shared in common to maximize opportunities for interaction and bonding.

Unlike residents of co-living arrangements who come together because of the lack of alternative affordable housing, residents of co-housing types are united by a shared interest of their long-term needs (Jarvis 2019), common social goals such as responding to the impacts of climate change (Chatterton 2016), and also a shared purpose of creating an intentional community (Sargisson 2018). To the extent that a co-housing community is made up of strangers who have decided to live together, the co-housing type also presumes some forms of shared collective governance—where community members meet, discuss, and debate on joint decisions (Sargisson 2018), and where agreements on matters pertaining to pets, shared food, and the use of shared space are consensually negotiated (Chatterton 2016). Beyond meetings, members also commit themselves to share tasks, burdens, and responsibilities in order to maintain the physical upkeep of the community as well as to satisfy their everyday needs (Sargisson 2018). For certain co-housing communities, possession pooling is also practiced, where large budget items that require occasional use are pooled and shared (Sargisson 2018). Importantly, the shared shaping of physical space is an integral aspect of the co-housing experience (Sargisson 2018: 153), which underpins both the source of relational elation but also of frustration and conflict. However, if Sennett (2012: 258) is correct, then engaging in this experience of dealing with conflict also creates the commitment necessary for sustaining this community.

In these ways and more, the co-housing can be deemed as a sharing configuration that enacts design by commoning. While professional designers may lay out the architectural groundwork for any co-housing configuration, and while the initial group of residents might have even decided on the overall layout of the community, much design work, which is shared, remains to be done. As Sargisson (2018) observes, the shared shaping of physical spaces in the co-housing configuration is a continuous task, where residents as commoners continuously shape their shared living spaces to correspond to their changing needs. Beyond the shared shaping of spaces, everyday decisions on major and mundane matters pertaining to the co-housing community have to be jointly discussed and debated. In turn, these decisions then shape the overall organization of the community in important ways.

In contrast to the 'design by capital' and the 'design by commoning' paradigm, we argue for the 'design by the systems approach' paradigm. We will discuss the components and the prospects of this paradigm in Chap. 4.

References

C. Alexander, *Notes on the Synthesis of Form* (Harvard University Press, Cambridge, MA, 1964)

C. Babb, C. Curtis, S. McLeod, The rise of shared work spaces: a disruption to urban planning policy? Urban Policy Res. (2018). https://doi.org/10.1080/08111146.2018.1476230

A. Badiou, in *Our Wound is Not So Recent* (Polity, Malden, MA, 2016)

N. Bhatia, A. Steinmuller, Spatial models for the domestic commons: communes, co-living and cooperatives. Archit. Des. Spec. Issue: Hous. Interv.: Archit. Towards Soc. Equity **88**(4), 120–127 (2018)

L. Böcker, T. Meelen, Sharing for people, planet or profit? analysing motivations for intended sharing economy participation. Environ Innov. Soc. Trans. **23**, 28–39 (2017)

P. Bresnihan, The more-than-human commons: from commons to commoning, in *Space, Power and the Commons: the Struggle for Alternative Futures*, ed. by S. Kirwan, L. Dawney, J. Brigstocke (Routledge, New York, NY, 2016), pp. 93–111

P. Chatterton, Building transitions to post-capitalist urban commons. Trans. Inst. Br. Geogr. **41**, 403–415 (2016)

L. Dawney, S. Kirwan, J. Brigstocke, Introduction: the promise of the commons, in *Space, Power and the Commons: the Struggle for Alternative Futures*, ed. by S. Kirwan, L. Dawney, J. Brigstocke (Routledge, New York, NY, 2016), pp. 1–27

M. De Angelis, Foreword, in *Common Space: the City as Commons*, ed. by S. Stavrides (Zed Books, London, UK, 2016), pp. xi–xiv

J. Dishotsky, Jon Dishotsky—Making city life accessible with affordable co-living solutions (n.d.), https://futureoflivingpodcast.com/tag/jon-dishotsky/
A. Escobar, *Designs for the Pluriverse: radical Interdependence, Autonomy, and the Makings of Worlds* (Duke University Press, Durham, NC, 2017)
K. Frenken, Editorial: sustainability perspectives on the sharing economy. Environ. Innov. Soc. Trans. **23**, 1–2 (2017)
B.M. Frischmann, *Infrastructure: the Social Value of Shared Resources* (Oxford University Press, Oxford, UK, 2012)
A. Gandini, The rise of coworking spaces: a literature review. Ephemer.: Theory Polit. Organ. **15**(1), 193–205 (2015)
R.E. Goodin, Institutions and their design, in *The Theory of Institutional Design*, ed. by R.E. Goodin (Cambridge University Press, New York, NY, 1996), pp. 1–53
G. Hardin, The tragedy of the commons. Science **162**, 1243–1248 (1968)
M. Hardt, A. Negri, *Commonwealth* (Belknap Press, Cambridge, MA, 2009)
M. Hardt, A. Negri, *Declaration* (Argo Navis Author Services, New York, NY, 2012)
D. Harvey, *The Enigma of Capital and the Crises of Capitalism* (Profile Books, London, UK, 2010)
S. Hodkinson, The new urban enclosures. City **16**(5), 500–518 (2012)
S. Holder, The largest co-living building in the world is coming to San Jose (2019), https://www.citylab.com/life/2019/06/cohousing-san-jose-room-for-rent-starcity-coliving-housing/590731/
J.D. Hunter, P. Nedelisky, *Science and the Good: the Tragic Quest for the Foundations of Morality* (Yale University Press, New Haven, CT, 2018)
T. Irwin, Transition design: a proposal for a new area of design practice, study, and research. Des. Cult. **7**(2), 229–246 (2015)
H. Jarvis, Sharing, togetherness and intentional degrowth. Prog. Hum. Geogr. **43**(2), 256–275 (2019). https://doi.org/10.1177/0309132517746519
E. Katrini, Sharing culture: on definitions, values, and emergence. Sociol. Rev. Monogr. **66**(2), 425–446 (2018)
Knight Frank, *Co-Living: Rent a Lifestyle* (2019), https://content.knightfrank.com/research/1004/documents/en/india-topical-reports-in-the-first-of-its-kind-industry-report-titled-co-living-rent-a-lifestyle-we-give-a-comprehensive-analysis-of-the-potential-for-rental-housing-in-india-6027.pdf
Knight Frank, in *Insights on Co-living: An Asia-Pacific perspective* (n.d.), https://content.knightfrank.com/research/1838/documents/en/co-living-insights-2019–6640.pdf
M. Kornberger, C. Borch, Introduction: urban commons, in *Urban Commons: rethinking the City*, ed. by C. Borch, M. Kornberger (Routledge, New York, NY, 2015), pp. 1–21
K. Krippendorf, *The Semantic Turn: a New Foundation for Design* (CRC Press, Boca Raton, FL, 2006)
M. Lazzarato, *The Making of the Indebted Man* (Semiotext(e), Los Angeles, CA, 2011)
I. Mariotti, C. Pacchi, S. Di Vita, Co-working spaces in Milan: location patterns and urban effects. J. Urban Technol. **24**(3), 47–66 (2017)
R. Martin, Real estate agency. in *The Art of Inequality: architecture, Housing, and Real Estate: a Provisional Report*, ed. by R. Martin, J. Moore, S. Schindler (The Temple Hoyne Buell Center for the Study of American Architecture, New York, NY, 2015), pp. 92–131
M. Mayer, *Civic City Cahier 1: social Movement in the (post-)neoliberal City* (Bedford Press, London, UK, 2010)
J. Merkel, Coworking and innovation, in *The Elgar Companion to Innovation and Knowledge Creation*, ed. by H. Bathelt, P. Cohendet, S. Henn, L. Simon (Edward Elgar, Cheltenham, UK, 2017), pp. 570–586
S. Olma, *In Defence of Serendipity: for a Radical Politics of Innovation* (Repeater Books, London, UK, 2016)
E. Ostrom, *Governing the Commons: the Evolution of Institutions for Collective Action* (Cambridge University Press, New York, NY, 2006)

G. Piccinno, Between the digital and the physical: reinventing the spaces to accommodate sharing services, in *Multidisciplinary Design of Sharing Services*, ed. by M. Bruglieri (Springer, Cham, Switzerland, 2018), pp. 41–51

J.P. Protzen, D.J. Harris, *The Universe of Design: Horst Rittel's Theories of Design and Planning* (Routledge, New York, NY, 2010)

C.M. Rose, The comedy of the commons: commerce, custom, and inherently public property. Univ. Chic. Law Rev. **53**(3), 711–781 (1986)

Y. Rydin, *The Future of Planning: beyond Growth Dependence* (Policy Press, Bristol, UK, 2013)

San Jose City Hall, Starcity Co-living Project (2019), https://www.sanjoseca.gov/your-government/department-directory/planning-building-code-enforcement/planning-division/environmental-planning/environmental-review/active-eirs/starcity-co-living-project

L. Sargisson, Swimming against the tide: collaborative housing and practices of sharing, in *Sharing Economies in Times of Crisis: practices, Politics and Possibilities*, ed. by A. Ince, S.M. Hall (Routledge, London, UK, 2018), pp. 145–149

S. Sassen, *Expulsions: brutality and Complexity in the Global Economy* (Harvard University Press, Cambridge, MA, 2014)

D. Saunders, *Arrival city: how the largest migration in history is reshaping our world* (Pantheon, New York, NY, 2010)

M. Schwarz, *A sustainist Lexicon: seven Entries to Recast the Future—Rethinking Design and Heritage* (Architectura & Natura Press, Amsterdam, The Netherlands, 2016)

R. Sennett, *Together: the Rituals, Pleasures and Politics of Cooperation* (Allen Lane, London, UK, 2012)

G. Standing, *The Precariat: the New Dangerous Class* (Bloomsbury Academic, New York, NY, 2011)

G. Standing, *Plunder of the Commons: a Manifesto for Sharing Public Wealth* (Penguin, New York, NY, 2019)

J. Wagner, D. Watch, *Innovation Spaces: the New Design of Work* (The Brookings Institution, Washington, DC, 2017)

Chapter 4
How to Design a Sharing System?

Shared systems can be defined as systems that provide a broad-based access and usage of a resource that is held in common. Broadly defined, many forms of public infrastructure are shared systems (Frischmann 2012). A motorway, a park, or a swimming pool are public amenities shared by different people. Other examples abound as well, and they include social security, education, and even defense. But being a shared system does not mean that it has been designed to motivate direct or purposeful sharing between people. In other words, a shared system has been primarily designed to optimize access or participation; it does not follow from this that individuals here would prefer to share, or are motivated to share, the common resource using more effective and efficient ways. And, although a shared system can increase social participation, and therefore also social returns (Frischmann 2012: 7), it has a tendency to suffer from overcrowding and overuse, which can lead to a reluctance to share this common resource. This is seen, for instance, in the preference to avoid traffic jams, or the preference for a more hygienic public pool. The counter-preference for not sharing is ever present even within a shared system. For this reason, a shared system is not yet a sharing system.

What then is a sharing system? As discussed in Chap. 1, a sharing system is the outcome of design. Following Boulding's (1956) seminal taxonomy of systems, a sharing system can be defined as a hybrid social system with the characteristics of an open system. Specifically, a sharing system is a social system composed of individuals acting in concert (Banathy 1996: 14). This action then enables the self-maintenance and also the self-reproduction of the sharing system. This sharing system is also a teleological, or goal-seeking entity (Churchman 1971), comprising other interconnected social and technological components interacting in an organized way to achieve this goal (Meadows 2008). Therefore, a sharing system is a formation made up of interconnected socio-technical components working toward the goal of reinforcing and reproducing more robust sharing behaviors and practices.

This goal can be further specified in the following three ways: First, a sharing system aims to motivate sharing behaviors by developing more effective or efficient,

J. K. H. Chan and Y. Zhang, *Sharing by Design*, SpringerBriefs in Applied Sciences and Technology, https://doi.org/10.1007/978-3-030-43569-1_4

or else more equitable and aesthetically pleasing ways to share. To formulate this differently, a sharing system promotes and also constitutes the agency for direct and purposeful sharing behaviors. Second, a sharing system, through the interaction of its components, provides clear and accountable relational procedures—rules, norms, and institutions—to support and sustain persistent sharing relations. On this, a sharing system aims to normalize sharing relations without calcifying them. Individuals would be accustomed to share in a sharing system, but they are neither expected, nor obliged, to share all the time. Third, a sharing system enables sharing with, and between, strangers. A sharing system aims to constantly evolve by integrating newcomers into a practice of sustained sharing behaviors and relations, and then to grow by expanding and evolving this practice in new social terrains.

Existing literature offers clues to further characterize the potential of this practice of sharing with strangers. For example, De Angelis (2017: 241) refers to this potential as a form of "boundary commoning"—where one practice of sharing joins with other practices resulting in a new and enlarged system of interdependencies. On the other hand, Katrini (2018: 431) describes this potential as the process where one taxonomical category of sharing expands and bleeds into other categories. However, for us, the practice of sharing with strangers represents what Carse (1986) calls, "infinite games": practices that are intentionally organized such that once initiated, they can facilitate, enable, and perpetuate their own persistence. To formulate this more concretely, sharing often leads to even more sharing, and through the design of the system, individuals are able to continue to share, and to extend the benefits of sharing to others.

4.1 The Design Principles of Elinor Ostrom's CPR Systems

If such a sharing system appears reasonably attractive, then how do we design it? And where should we even begin? Here, Ostrom's (2006) seminal work on the governance of common-pool resources (CPR) appears to offer a starting point. Ostrom (2006: 90) refers to the principles behind successful commons as 'design principles'. Furthermore, Ostrom conjectures that these design principles are applied in different successful CPR systems. But exactly what do these design principles mean? Are design principles merely a formal recognition for institutional design (see Goodin 1996)—defined as structuring principles necessary for guiding institutional changes and improvements for the CPR system? Or are these design principles the necessary conditions for successful CPR systems to be true?

Ostrom observes that despite different contexts, successful CPR systems tend to share a similar set of settings. First, Ostrom (2006: 88) emphasizes that uncertain and complex environment prompts the constitution of the CPR. Where there is environmental uncertainty in resource availability and where dealing with this uncertainty demands complex socio-technical cooperation, individuals tend to share the CPR. Second, the population involved in CPR management has tended to remain stable across time. These people have shared a past, and are expected to share a future (Ostrom 2006: 88), which suggests that they can expect continual cooperation to produce continuous and mutual benefits far into the future. In other words, there is little reason to defect from cooperating on the management of the CPR. Third, Ostrom (2006: 89) notes "the sheer perseverance manifested in these resource systems and institutions". In spite of an uncertain and complex operation environment, and in spite of all the opportunistic ways in which individuals could defect from cooperation, these commons have endured across time. Given these similarities, the primary objective is then to explain the sustainability and robustness of successful CPR systems (Ostrom 2006: 89).

And because the successful CPR is a *given* empirical fact, Ostrom could then proceed to infer the principles that can begin to explain this success. Ostrom (2006: 90) outlines eight design principles that might explain this "sheer perseverance" of successful commons. They are namely, (i) Clearly defined boundaries; (ii) Congruence between appropriation and provision rules and local conditions; (iii) Collective-choice arrangements; (iv) Monitoring; (v) Graduated sanctions; (vi) Conflict-resolution mechanisms; (vii) Minimal recognition of rights to organize, and for CPRs that belong to larger systems; and (viii) Nested enterprises. Therefore, the process is inductive and the aim has always been about explanatory power of the theory. Following this, a subsequent stage entails testing this theory deductively on other specific cases of CPR (see Cox et al. 2010). Therefore, we suggest that Ostrom's approach is primarily scientific and not, as the phrase 'design principles' appears to suggest, prescriptive or design-oriented. At best, Ostrom's principles here suggest that design activities could have occurred, but they fall short of demonstrating just how the design process might have been applied to attain a successful CPR system.

4.2 Justifying the Systems Approach for Sharing by Design

In contrast, our approach is design-oriented and prescriptive. To paraphrase Banathy (1996: 19), our approach perceives the sharing system as a system of interconnected, interdependent, and interacting components, and aims to create a design solution as a sharing system made up of interconnected, interdependent, interacting, and internally consistent solution ideas. It is therefore concerned first and foremost with this design question: 'how to design a sharing system?'. But to answer this design question, it is necessary to first address the following second-order epistemological question: 'what are the components, parameters, and considerations that must be taken into account when designing a sharing system?'. In other words, what is the knowledge necessary for the design of sharing systems? This is a question that we seek to address in this chapter. And diametrically different from Ostrom's (2006) work, we do not, and cannot, take the success of a sharing system as a given empirical fact, and then deduce the principles that must be present in order for the successful sharing system to be true. In other words, unlike Ostrom's approach, we are unable to deduce functional principles as hypothesized design principles from sharing systems that exist and are, moreover, successful. At best, Ostrom's principles are instructive, but they do not constitute the basis of our systems approach. Instead, we posit the success of a sharing system as the design goal, which the design process then attempts to approximate. If the objective of Ostrom's work is to discover the general principles behind the success of CPR systems, then our aim is to design a successful sharing system. To formulate this differently, we aim to postulate the knowledge and method for designing a sharing system.

Importantly, we have modeled our design method after Churchman's (1968) systems approach, which is further informed by Meadow's (2008) more recent work in systems thinking. But what makes our design method modeled after the systems approach relevant for sharing by design? Why, exactly, should the systems approach be privileged over other major design approaches—for example, the rational-computational design approach (Simon 1996), the increasingly popular participatory or co-design approach (Manzini 2015), and the dialogic-reflective design approach (Schön 1990)—to name just three major approaches in design? After all, to define something as a system is a design choice (Churchman 1971: 42), and this choice has to be justified.

We argue that when compared to these three major approaches, the systems approach offers the best fit for the design challenges peculiar to sharing systems. This best fit can be justified in the following two ways. First, and relative to other design approaches, the systems approach permits a clear visualization of important components within a sharing system. The systems approach then aims to represent these components and their dynamic relationships abstractly, but nevertheless clearly, to the designers and other participants. The transparency of the systems-driven design process then can facilitate more robust discussions among the participants. Second, these system components are represented as discrete categories; they are neither ordered nor weighed in any pre-conceived way. This can permit the skillful system designer to design various alternative sharing systems by using the same building blocks: varying their relations, ordering or weightage, and correspondingly, also their interactions and feedback. In turn, this process then produces different outputs and system performances. The ability to create alternative systems is deemed as one of the primary strengths of the systems approach (see Churchman 1971: 46). Because the context of sharing systems can vary, this ability to vary the design of sharing systems systematically according to different contexts is therefore a key advantage of this design approach.

To further justify the systems approach, we will briefly compare it to the three major approaches in design, which are namely, the rational-computational approach, the participatory and co-design approach, and finally, the dialogic-reflective approach. The aim here is not only to argue how each of these three approaches is limited for designing the sharing system, but also to explain why these limitations are particularly debilitative for the design goal of a sharing system.

To begin, the rational-computational approach first aims to define and structure the design problem, and then to search for the optimal solution within the parametrically bounded solution space of the defined problem. The rational-computational approach is rooted in the science of Operations Research (Churchman 1979: 16; Simon 1996: 27). But beginning with Churchman's (1968) later work on the systems approach—especially with his unfinished meditation on how to properly secure systemic improvements (see Ulrich 1994)—Churchman started to depart from the paradigm of Operations Research. Gradually, Churchman realized that the measure of optimality alone is insufficient for the weightier question of system sustainability. In other words, while the rational-computational approach excels in seeking out the optimal design solution by increasingly refining its bounding parameters, it could be largely agnostic to the sacrifices that might have to be made to attain this optimality (Churchman 1968: 4)—sacrifices that could unsettle the entire system. Furthermore, for the optimal solution to remain in this state, it has to presume different social relations that have not been taken into account in this design approach. The warrant of these social relations has been corroborated by Ostrom's (2006: 94) design principle of monitoring, where in order for the CPR to remain in an optimal state, the community has to actively audit its conditions and monitor the behaviors of resource appropriators. For these reasons, the rational-computational design approach remains inadequate for the design of a socially complex sharing system.

Second, participatory and co-design approaches are also relatively limited for the design of sharing systems. Invariably, sharing systems are participatory in nature, and may even practice some degree of self-design, where participants try to shape their sharing systems together. Although the participatory approach for generating design knowledge is deeply meaningful, it is, however, also highly contingent and uncertain. An instance of a successful participatory design in one context may not be replicated in other contexts. While the participatory approach can be systematic, it cannot be fully rationalized. Designers employing the participatory approach may never come to formulate the necessary constituents or principles underlying any instance of successful participatory design. In contrast, the systems approach tries to outline as many of these constituents and principles as possible in order to build a robust model—a complex hypothesis—of successful sharing systems. This model is especially important when a sharing system fails to attain its goals. Based on this model, the designers could at least conjecture on what went wrong.

On the other hand, co-design begins with a powerful default starting point, which is the parity of politics in design. In co-design, everyone is an equal designer to everyone else. Admittedly, this is the primary strength of the co-design approach. But what is often left unsaid is its chief weakness: if everyone has an equal right to make design decisions, then exactly how should a decision be made? And perhaps more importantly, by what means? To these important questions, the co-design approach does not offer any immediate answers. However, Marx (1978: 364) offers a clue: "between equal rights force decide". Even if one does not concede to Marx's pessimistic projection, one may have to concede that although co-design begins on a promising note, it is nevertheless an underconstrained design approach that bears the risk of an uncertain ending—one that may eventually be mired in conflict and strife. On this, while the systems approach does not, and cannot, specify the parity of design politics in advance, this approach at least offers some transparency on the anticipated mechanisms of the model underlying the sharing system. In turn, this clear visualization of the sharing system can facilitate further negotiation and also cooperation on how to attain the goal of this system.

Finally, Schön's (1990) dialogic-reflective design approach stands farthest from the systems approach. For Schön, nearly every design parameter—even the design problem—can be unsettled or overturned by learning in the design process. In Schön's (1990: 136–137) approach, designers 'converse' with the materials of the design situation and they learn from this conversation. Through this learning, the structure of the design problem also evolves. Importantly, this suggests the unsettling possibility that the designers might even learn that design is futile and subsequently, to relinquish design completely. In other words, while Schön's design approach is powerful, it cannot assume the persistence of design and at the same time, it has to concede to the remote but possible cessation of design activities. But conceived more positively, designers relying on Schön's approach are able to transform the definition of the design problem in ways that cannot be predicted in advance, which paradoxically renders this approach underdetermined yet powerful. In this way, Schön's approach is thoroughly rigged with uncertainties but it is in the same vein, also laden with latent potentialities. The systems approach cannot hope to match the full power of Schön's approach. Yet by the same token, the systems approach also does not need to concede to the radical limitations found in Schön's approach. In Schön's design approach: when the dialogue between designers and the design situation ceases, the design process also terminates. This is a fatal possibility that sharing by design cannot afford. For the sharing system to persist, new dialogues must be constantly invigorated in place of old ones that have ceased. To avoid this possibility, Schön's dialogic-reflective approach may be instructive, but not decisive, for sharing by design.

4.3 Specifying the Components of the Sharing System

At this point, we specify the components of the sharing system. Importantly, specifying each of these components here is not a substitute for demonstrating the design of a sharing system. To demonstrate how to design a sharing system, we will simulate design by applying these components in the hypothetical design of a waste-to-energy sharing system in the conclusion of this chapter.

Based on our study of Churchman's (1971) systems theory, we propose at least eight general components present in any sufficiently complex sharing system. They are namely, (i) The environment; (ii) Goals; (iii) Guarantors; (iv) Performance measures; (v) Stakeholders; (vi) Resources; (vii) The social forces: major ones include culture, politics and ethics; (viii) Enemies (of the system). Two key caveats are in order here.

First, we have selected to present these components as discrete categories. However, in any organized system, these are often overlapping and also interdependent components. For instance, the external environment of the sharing system constitutes many, but not all, of the social forces within the sharing system. Similarly, the composition of the stakeholders shapes the nature of goals in the sharing system, and in tandem the stakeholders themselves are moderated by the environment in significant ways. In turn, these impact the nature or metric of the performance measures. These specific interactions and interdependencies cannot be prefigured in any sufficient resolution because every specific sharing system is different in some significant ways. The most accurate conceptual model we can offer is a general model comprising of these components independently described. This said, where applicable, we will still highlight any possible interactions between these components.

Second, the seventh component of 'the social forces' is a metaphorical bundle of several social categories. This component serves as an artificial cap on the innumerable social forces that can modulate the sharing system. We have relied on culture, politics and ethics to represent this artificial cap while admitting that there are many other social categories that are missed by these general headings. The prudent systems designer may want to first acknowledge this as a general category, and then vary the emphasis or attention to these three or other important social forces at work in his or her design of a specific sharing system.

4.4 The Environment (of the Sharing System)

All systems have their corresponding environment, which is posited as an external reality that resides beyond the boundary of the system. This environment can vary in its stability, its complexity, and its threat (Scott 1992: 143), and can impact the system in different ways. For example, Ostrom's (2006) work suggests how the CPR system is a response to an environment characterized by complexity and uncertainty. But a sharing system can also respond to its environment proactively. Katrini (2019) suggests that a sharing system can leverage on the potentials of a city by making itself more publicly visible near important thoroughfares, or by choosing to situate itself close to other existing amenities that can reinforce sharing. Conversely, a sharing system could begin to undermine itself when it neglects its environment, for instance, when shared bikes clutter the sidewalk after their use (Chan and Zhang 2018). In describing the environment this way, it is framed as a component that a systems designer may wish to first acknowledge, and then subsequently, to discover how to work with, rather than against, it.

The environment is important in systems thinking. As a matter of fact, Churchman (1979: 4) has reserved a special fallacy regarding it as, 'the environmental fallacy'. The environmental fallacy entails ignoring the external reality when solving a design problem. Consider the social nuisance caused by bike-sharing programs. The designers of the bike-sharing system have been highly effective in optimizing the availability of bikes to riders who need them. However, this optimality depends in part on how widely available these shared bikes are, which in turn tends to mean that they could be found nearly everywhere—but often found parked in irresponsible ways and causing a great deal of public angst (Chan and Zhang 2018). In this way, even when a sharing system made up of these shared bikes and riders is steadily becoming more efficient (i.e. a state where bikes are easily available to riders who need them), this efficiency, however, begins to undermine the effectiveness of the bike-sharing system when pedestrians are repelled by sidewalks littered with bikes. In other words, the optimization of bike-sharing here has largely ignored its impacts on the environment—and the costs of this ignorance are tremendous.

Following this, can the designer then try to design—that is, deliberately control for—the environment of the sharing system? In other words, can the designer extend the boundary of the sharing system to include other systems that were once regarded as part of the environment, and in doing so, reinforce the sharing system? There has been a long line of thinking on a system enacting its environment in organizational science (Scott 1992: 141), which entails gaining control on the environment in order to reduce the system's own uncertainty. A vivid example of this took shape in the R-Urban project in Colombes, north-west of Paris. This was an attempt in creating an ecology of local practices (Petrescu et al. 2016: 722), where sharing activities expanded beyond any one specific urban site to create an enlarged and interdependent ecology of sharing practices, which then became their own immediate environment. In a similar vein, researchers and designers have started to conjecture on how co-working as a sharing practice can begin to transform surrounding urban areas (Armondi and Di Vita 2017; Katrini 2019).

4.5 The Goal (of the Sharing System)

The goal is the raison d'être of the system: it is the objective that a system works to attain. On this, goals can be further distinguished into intermediate goals and the final goal. While the former has to be attained in order to reach the final goal, the latter is the summative purpose of the system. In this way, successive or persistent attainment of different intermediate goals improves the odds of attaining the final goal. And also for this reason, it is important that intermediate goals do not conflict, but rather, reinforce the final goal. Even so, consistent intermediate goals can nevertheless conflict. For example, in a sharing system, if the final goal is the long-term persistence of this sharing formation, then the intermediate goals are likely to entail building trust between present participants and keeping this sharing community open to newcomers. However, these two intermediate goals can conflict, when newcomers do not cohere well with the present participants, and this conflict may then end up destabilizing the final goal. It is therefore important to evaluate the independent impacts of each intermediate goal, and also their interplay in relation to the final goal of the sharing system.

The final goal sets the overall direction of the system. In this way, the final goal overrides, but also regulates, the intermediate goals (see Meadows 2008: 162). This means that it is possible to assess the appropriateness of intermediate goals using the final goal as a benchmark: intermediate goals that are consistent to the final goal are admitted, while those that conflict with the final goal are rejected. The final goal then guides and shapes the system. But how is this final goal benchmarked? And by what standards, rules, or imperatives? It is often challenging to assess the appropriateness of the final goal if only because the designer will require a higher goal to do so—a goal that is not always clear or easily defined at the outset. For this reason, Meadows (2008: 138) argues that systems can easily seek the wrong final goal. For instance, if the final goal of a sharing system is defined as more efficient sharing, and efficiency is measured by increasing the number of participants while conserving the shared resources, then the system will operate to produce as many participants as possible. Because the goal is now measured by the number of participants, this system may or may not produce the behaviors and relations that are integral to an enduring sharing system.

Is this then merely about the problem of choosing an appropriate metric? Or does the problem still pertain to the challenges of defining a final goal? We suggest that both the proper selection of a measuring metric as well as the appropriate definition of a final goal are important considerations. But we emphasize that the appropriate definition of a final goal ought to be prioritized. This is because the choice or the definition of the final goal has to precede its own metric. Furthermore, the question on the choice of an appropriate metric is only meaningful if this final goal is also an ethical one. To apply an appropriate metric to an immoral goal can only result in an absurdly unethical outcome, for instance, when a designer measures the performance of a machine gun by projecting how many people it can kill in a minute. In line with Churchman's (1979) definition of ethics as the theory of appropriate goals of a system, we propose ethics as a possible approach for defining and assessing final goals. We will further discuss the role of ethics in sharing systems in Chap. 5.

4.6 The Guarantors (of the Sharing System)

Observing that systems are fragile, and that they often also fail to secure overall improvements, Churchman (1968) wonders if there is a component that operates to secure, ensure, and guarantee a system's performance and persistence. What are entities that can guarantee the persistence of the system (Churchman 1979: 99)? And to this we add: How can we design these entities into the sharing system once they have been identified?

For a start, it is important to offer more vivid examples of the guarantors. In one story from the Judeo-Christian tradition, the Israelites were wandering in the desert. There, they were told to gather Manna (i.e. the bread falling from heaven) for food. As the story unfolds, some gathered more, and some less—yet when they measured what each had gathered, those who gathered much had nothing left over, and those who gathered little had no lack (Exodus 16: 16–18). It appears that there was an inexplicable guarantor at work to ensure sufficiency—that despite the varying motivations and efforts of the Israelites, everyone had enough. And more empirically, experimenters pay a show-up fee to subjects that bother to report to the laboratory. This show-up fee policy is then a guarantor that ensures that everyone who travels to the laboratory will receive compensation. This guarantor also ensures that the attempt to recruit subjects would not be futile. In a similar vein, a small penalty serves as the guarantor for ensuring that library patrons continue to respect the 'no late returns' rule even when lateness is negligible or incidental. This guarantor ensures that more, rather than fewer, books are returned to the library when they are due. From the inexplicable to the mundane, guarantors pervade everyday systems. Guarantors ensure that feedback loops work in the way that they are designed to work; they regulate the operation of the system in the direction of its goal. Guarantors are the failsafe of systems.

In spite of critical importance of the guarantor, acknowledgment of this component in systems thinking—except for Churchman's own (1968)—has been rare. For this reason, specific discussions of the guarantor in the context of design are even rarer. The guarantor is truly an undertheorized yet important aspect of systems design. Here, we extend the discussion on the guarantor within the frame of sharing by design with two further points.

First, as a concept, the guarantor appears vulnerable to infinite regression: what guarantees the guarantor—and on it goes in an infinite regress? Churchman (1979: 98) himself was keenly aware of this possibility and he posited the ultimate terminal of this regress as 'God', or the 'Guarantor of Destiny'. However, for the purpose of sharing by design, theology is neither immediately relevant nor practically useful. It is sufficient, by the logic of satisfice (Simon 1996), to seek out only the good enough. And it is therefore good enough to discover or to define the guarantors that are relevant in the sharing system, and then seek to implement them in practice without the need to check for further regress. As it stands, the mundane examples discussed all reveal commonsensical answers to the question of 'what or who guarantees the guarantor'. For the show-up fee, at the very least, is research funding; and as for the late penalty, a library committee willing to enforce this policy. Therefore, insofar as the guarantor does not rely on an inexplicable and indeterminable source, we suggest that the possibility of infinite regress is not immediately fatal for **the** idea of the guarantor.

Second, the guarantor can be designed. Consider the coin-operated locker found in many museums, which is also a shared system. To access one of these lockers, one first deposits a coin, which triggers the release of the key to the locker. To retrieve one's belongings later on, one uses this key to unlock the locker, which then releases the coin back to the user. As an assembly, this coin-operated lock serves as a guarantor to ensure some degree of responsibility for the key, which is an easily displaced small object. The act of having to deposit a coin before using the locker registers a small but conscious reminder to be more careful with this key. As a result of this design feature, these locks always return to their default setting ready to be used by the next user. It is possible to suggest that this design feature as guarantor then guarantees the continual operability of these lockers.

4.7 The Performance Measures (of the Sharing System)

The designers of the sharing system will need to know how their design is performing. Is the system performing according to the way that it has been designed? Is the system underperforming, or perhaps, exceeding the designers' expectation? To answer these questions will require the designers to establish some kind of measures to evaluate the performance of the system. Here, it is possible to distinguish between objective and subjective measures. The former are objectively true or applicable measures for all sharing systems, while the latter involve measures that may vary according to different stakeholders, or *clients*—individuals or social groups that the sharing system serve.

One objective measure is a performance measure that has been derived from the goal of the sharing system. As discussed, if the goal is simply stated as more efficient sharing, then one objective measure entails evaluating the performance of the system based on the number of clients it is able to serve. An increasing number of clients, at least up to a certain threshold, would mean that the sharing system is performing well. Conversely, a decreasing number of clients is likely to imply that the sharing system is underperforming. And if the goal of the sharing system is stated as self-perpetuation, then an objective measure at least entails finding out if outgoing clients have been balanced by incoming clients.

Even so, meeting any one objective performance measure does not guarantee the success of the sharing system. This is because 'success' is a subjective performance measure. Does success mean achieving most, if not all, of the goals of this sharing system? Or does success refer to exceeding these goals—for instance, that the sharing system has substantively improved an even larger system that it is trying to influence? Or does success narrowly imply sidestepping major systemic pitfalls or avoiding failures? Clearly, how success is defined will at least depend on the environmental conditions of the sharing system, and also on the preferences and expectations of the clients. In a largely favorable environment, a highly motivated client body is not likely to be only satisfied by the sharing system meeting its stipulated goals, it has to exceed them. Conversely, in a highly constrained and uncertain environment, even a somewhat enthusiastic client body has to temper its expectation. For these reasons, the goal of the sharing system cannot completely determine the performance measures of this system. Careful considerations of the clients' expectations in relation to the environmental conditions are also important.

Ineluctably, the sharing system also has a peculiar characteristic rarely found elsewhere. In a sharing system, the clients not only bear the standard of the performance measures (Churchman 1971: 43), but they are also involved in motivating the future performance of the system itself. In other words, the clients are the embodiment of sharing practices that in turn determine the subsequent performance of the sharing system. In this way, the sharing system comes close to an autopoietic system (Mingers 1995), where clients produce sharing activities and practices, which in turn drive the sharing system that facilitates or enables sharing, or else reinforce these clients in some significant ways (see Escobar 2017). In a sharing system, the clients embody the performance of the system.

4.8 The Stakeholders (of the Sharing System)

As aforementioned, the clients are the stakeholders of the sharing system. These clients include the immediate individuals or social groups that participate in the sharing system, and also a group of unknown users that rely on this sharing system occasionally, but are yet formal members of this system. Other important stakeholders include the designers of the sharing system, who conceptualize and define the specific inter-relationships within this system, and also municipal authorities that have immense discretion in supporting or even repelling this sharing system (see Petrescu et al. 2016). A far-seeing designer of the sharing system will also consider the real-estate developer, who sees the potential in the rising valuation of the adjacent urban sites by sharing activities, as a potential stakeholder in the system.

Ostrom's (2006) work suggests that stakeholders are central in a social system—yet they can be the most underdetermined component in the system. To better predict and to regulate their behaviors, an intricate set of rules pertaining to monitoring, sanctions and conflict-resolution has to be devised in a CPR system. Even so, a sharing system will still face the challenge of behavioral regulation. However, urban sharing systems are complicated by the reality that they have to admit a far wider range of clients than Ostrom's CPR system—clients who may not readily concede to mutual monitoring, or who can easily evade sanctions by exiting the sharing system (Hirschman 1970). In the absence of social solidarity, the prudent designer of the sharing system will have to find ways to build and reinforce bridging social capital between plural clients in the system.

4.9 The Resources (of the Sharing System)

Shared resources constitute another important component of the sharing system. Resources are important because not only are they central to the sharing relations, but their nature can also constrain the form and the process that sharing takes. There are at least three distinctive categories of resources (Chan and Zhang 2018).

First, there are resources that are tangible, subtractable, divisible, and also rivalrous in nature. These resources tend to take the form of discrete and usable objects. The different tools available in a shared tool library are examples of such resources. Sharing this type of resources tends to imply division: where if more of these objects are shared, then there are fewer left to share. Their subtractable and rivalrous quality also means that if one individual is using a resource, then it cannot be shared with another individual, and so sharing here adheres to a form of zero-sum sharing (John 2017). Because of this rivalrous nature, these resources can also be perceived as scarce goods. For these reasons, the sharing of tangible and rivalrous resources would do well to multiply the range of these resources into a cornucopia of shareable goods, for example, in the form of a well-stocked tool library. In this way, sharing activities do not need to cease when certain resources are in use; there are always some other resources that can be shared in the meantime. And because such resources can be perceived to be scarce, distributive rules become critical for this sharing system: is sharing operated on the priority of 'first come first serve', or the rule of 'sharing out to the greatest need'? The practice of sharing tangible and subtractable resources therefore renders the notion of fairness visible, salient, and perhaps, even contentious.

Second, there are resources that are intangible, non-subtractable, non-excludable, and indivisible. These resources are not depleted by an increase in either the frequency or the magnitude of sharing. For example, sharing news, information, or knowledge are clear instances of such resources. Kornberger and Borch (2015) rightly claim that using these resources through more frequent sharing does not deplete but rather increases their value. This resonates with the insights of Rose's (1986) 'comedy of the commons', where free and open access to certain shareable resources has the surprising effects of amplifying their values. For this reason, this type of resources tends toward non-zero-sum sharing, where the total value of sharing is neither fixed nor capped. Importantly, this particular quality can enable highly efficient ways of sharing, where sharing amplifies the original value of the resource. Instead of a mostly dyadic relation observed in the sharing of tangible resources (i.e. between a provider of the resource and the user), here, the sharing of an intangible resource can mean its transmission to many users at once, and the amplification of the original resource—for example, when a teacher shares her knowledge with many students, or vice versa, when a student accepts this knowledge and shares a reworked version with the entire class.

Third, there are resources that straddle between tangible and intangible resources. On one level, these third type of resources are subtractable and rivalrous; yet in sharing them, new non-subtractable and indivisible resources as new experiences and knowledge are also produced and shared. On another level, sharing these resources can create new relations, rapport and bonding social capital, which lead to the formation of a creative agency that may transform the very sharing system itself. One typical example of this third type of resource is the physical space. While sharing more of a limited space usually means having less to share, however having more participants share the same space can also lead to a new agency for reshaping the space together (see Sargisson 2018: 153). In other words, the law governing the sharing of subtractable and rivalrous resources does not appear to entirely apply to the practice of sharing spaces. Quite the contrary, this practice may instead provoke a creative agency that can reshape the limitations of the resource in ways that mitigate their worst outcomes.

4.10 The Social Forces (of the Sharing System)

The social forces can be conceptualized as a mysterious black box in the sharing system: the designer of the sharing system knows that these forces exist but they can neither be systematically enumerated nor their operations accurately known. Here, it is important to note that the use of 'forces' is metaphorical; they do not imply Newtonian physical forces. These 'forces' are actually closer to what Durkheim (1982) refers to as 'social facts': ideas, beliefs, and categories that are external to an individual, but can constrain this individual's actions. One example of a relevant social fact in the sharing system is the sharing culture (Katrini 2018; Light and Miskelly 2015). In a sharing system that maintains a culture of demanding participants to hold each other accountable on what they share, and also how they share, this system will operate in a starkly different manner from one that merely conceives sharing as a form of transaction (Light and Miskelly 2015: 59). Beyond culture, another example of a relevant social fact is the politics of the sharing system. A sharing system where participants not only share resources, but also share an equal power to decide how to run this system (Stavrides 2016), will function differently from a system where management is centralized.

Here, it is important to distinguish between social forces that form part of the environment of the system, and a cohesive culture that can arise through social interactions within the system. While the former is external to the sharing system, the latter emerges from the system. For external social forces, consider the example of bike-sharing again (Chan and Zhang 2018). Bike-sharing systems operate differently in different societies not only because of differences in physical geography but also because of differences in the cultural perception of sharing bikes. In certain societies, for instance, people can accept some cluttering of bikes on the sidewalk. In other words, there are people willing to trade-off some orderliness for more practical conveniences (i.e. to be able to find a bike when one needs it). On the other hand, in other societies, no amount of convenience is appealing enough; orderliness has to be prioritized. This has led to the creation of designated bikes parking zones (Channelnewsasia 2017), but which can undermine the appeal of convenience for this shared mobility system. While these external social forces do not determine the outcomes of the sharing system, they can nevertheless shape this system in important ways.

However, different components can also interact within a sharing system to create an internal culture. Can a designer then purposefully arrange the different components of a sharing system such that they interact to produce a culture that then reinforces sharing behaviors? Consider the example of an autonomous community of practice. An autonomous community that practices the design of itself also endogenously produces its own norms (Escobar 2017). In turn, these norms then reinforce the community. In this case, autonomy is then a guarantor for the creation of the internal social forces that can draw the community closer. Or consider the possibility of accelerating feedback between action and consequences in order to reinforce ethical behaviors. Here, Skinner (2002: 172–173) suggests that ethics is concerned with bringing the remoter consequences of behavior into play. A sharing system that permits clients to monitor each other (Ostrom 2006), when compared to one that recognizes anti-social behaviors only after a long delay, is likely to reinforce the ethic of accountability within the system. In this way, Escobar's (2017) reminder is timely: design can 'hardwire' particular kinds of politics into bodies, spaces and objects. If this is true, then designers can arrange the interactions between the components to induce the necessary social forces for reinforcing the durability of a sharing system. In Chap. 5, we will focus on ethics, which is one of these social forces.

4.11 The Enemies (of the Sharing System)

'The enemy' is Churchman's exclusive analytic invention in systems theory. Churchman (1979: 13) defines 'the enemy' as a consortium of opposites: an enemy is both distrusted and admired; an enemy is both respected and feared. The enemy is, therefore, a legitimate adversary that threatens the system. This enemy may or may not be real; its facticity is quite beside the point. The primary purpose of the enemy is to function as a dialectical category: the enemy represents an opposition to the purpose or the philosophy of a sharing system. It is furthermore an embodiment of different interest positions that could be advanced had this sharing system not existed. The enemy, in other words, has every incentive to eliminate a sharing system. To conceive this enemy ahead of time, and to anticipate its many possible oppositions, is no different from designing the defenses of the sharing system.

To begin, it is first important to identify the potential enemies of the sharing system. Ostrom (2006) distinguishes the CPR system from its primary contenders, which are namely the state or centralized management of resources, and the privatized, market-driven management of resources. Based on this, it is possible to suggest that capitalism is one potential enemy of the sharing system. Referring back to Chap. 3, we suggest the possibility of a market-driven and capital-led approach to designing for sharing—with its merits but also critical shortcomings. Other researchers have discovered different assaults from this particular enemy. For instance, Ciaffi (2019) discovers that the lack of stable financing could undermine the stability of urban sharing systems. In the same vein, Huron (2015) documents the persisting appeal of 'selling-out' integral parts of a sharing system to real-estate interest in a capitalist landscape. For others, the enemy could come in the form of municipal politics (Petrescu et al. 2016), where municipal support for the sharing system was retracted once political offices changed. In the face of these overwhelming enemies, it is too easy for the designer of the sharing system to grow disheartened.

Once potential enemies have been identified, the designers have the following choices. The designers can design for specific defenses by anticipating the enemy's every possible move, or the designers can design for an overall systemic resilience, which can cushion the sharing system against the assaults of the enemy. These choices are not mutually exclusive; the designers can choose any appropriate combination between these two choices. However, it is important to note that each of these choices entails very different design strategies. While the former entails creating counters to specific and anticipated assaults by the enemy, the latter involves boosting overall systemic constitution in order to absorb the different and somewhat unpredictable assaults by the enemy. To formulate this differently, the former presupposes the capability to read the enemy well, and aims to reorder the sharing system to deflect the enemy's specific assaults. On the other hand, the latter is agnostic to the enemy's behavior, and instead aims to improve the resilience of the sharing system.

4.12 A Brief Simulation of Designing a Waste-to-Energy Sharing System

To demonstrate how to design a sharing system, we will briefly simulate the design of a waste-to-energy sharing system in the city-state of Singapore. For a more detailed analysis on the design of sharing systems in the context of an architectural design studio, please see Chap. 6.

Cities today are burdened with the worsening problem of waste, and this problem is especially acute in land-scarce Singapore. Presently, the off-shore Semakau landfill is the only waste storage facility for Singapore, and this landfill handles incinerated ash and non-incinerable waste shipped from mainland Singapore on a daily basis. This landfill is also anticipated to reach its full capacity before the original scheduled year in 2045 (Chan 2016). For this reason, effective and creative solutions are being sought to reduce the volume of waste sent to the Semakau landfill—to delay the moment of full capacity for as long as possible. One solution, which targets the relatively large volume of food waste produced in Singapore, is to reduce and reuse food waste. Specifically, this solution aims to produce biogas by combining water sludge and food waste in an anaerobic digester (Teh 2019). Through this solution, the recovery of resources from food wastes can be maximized on the one hand, and on the other hand, this contributes a small but tangible step toward Singapore's energy self-sufficiency (Teh 2019). Presently, the municipal authorities are looking at scale-up operations based on this solution. However, the designer of a sharing system can conceive a 'scaled-down' version of this solution as the basis of energy sharing for a neighborhood. As a matter of fact, precedents of these 'scaled-down' approaches already exist in Singapore, where food waste are emptied into eco-digesters, which decompose waste into potable water that is in turn used to clean the cooked food centers that produce this waste on a daily basis (Boh 2017).

In these ways, the environment is propitious for the success of this waste-to-energy sharing system. First, this is an environment where public perceptions on waste are rapidly changing. Not only is a 'business-as-usual' policy for food waste perceived to be problematic, but also increasingly, waste is also perceived as a valuable resource (Zapata and Campos 2015: 106). Moreover, suitable technologies already exist to reduce and reuse food waste on a permanent basis. Importantly, the sustainable goal of sharing systems (Cohen and Munoz 2016), also aligns with this environment. If a community can converge to create an energy co-op (Cohen and Munoz 2016: 94), then this foundation will form the basis of other modes of co-production. Following this, new resources generated from new modes of co-production, now propelled by the waste-to-energy power, then go on to guarantee the sustainability and persistence of this sharing system.

Concerning these possibilities, the designers must now consider where to locate the sharing system. In Singapore at least, people congregate to dine in, and to socialize, in cooked food centers—or hawker centers—on a daily basis. As aforementioned, a few of these hawker centers already operate eco-digesters, and so these hawker centers form the natural starting points of this sharing system. Following this, the designers must now find out how to galvanize a critical mass of stakeholders that can contribute their food waste to these eco-digesters. Would these stakeholders then include individuals and families living near to the hawker center, where they are incentivized with credits that off-set their monthly utilities bill whenever they 'share' their food waste? Or should stakeholders also include eco-entrepreneurs who can bring in substantial amounts of food waste elsewhere, and then leverage on the power produced to generate new marketable products and services? Whichever is the choice, expanding the boundary of the sharing system means increasing the number of stakeholders, and it may be important to consolidate the fledgling system with a sufficient number of key stakeholders first.

In tandem, the designers have to start to define the performance measures of this waste-to-energy sharing system. Is the key performance measured by how much waste has been used to generate power? Or does performance also include the quantities of biogas produced for electricity generation? As discussed, one way to define an objective performance measure is to derive this from the final goal of the system. If so, then an objective performance measure is the amount of reusable energy generated by this sharing system, and further on, also the income generated from the co-production enabled by this new energy source.

As the organization of this sharing system gradually takes shape, the designers would do well now to consider both the guarantors and the enemies of this system. One possible guarantor is the perpetual generation of food waste, which is the primary input into the system. As a side-effect of cooked food preparation, production, and consumption, a certain amount of food waste can be always guaranteed. However, this guarantor can paradoxically become its own enemy—where a highly efficient waste-to-energy conversion process now redeems, and perhaps even encourages, the behavior of wasting more usable food. It is therefore important for the designers to impose other guarantors that can limit this disposition, for instance, by designing a culture of zero waste or by imposing an ethic of less waste so that the probabilities for these paradoxes of waste-to-energy generation can be diminished.

In sum, this simulation of the design of a waste-to-energy sharing system recalls the following broad design strategies (however, not necessarily in the following sequence):

(i) Defining the environment, or the problem scenario.
(ii) Identifying the opportunities for a sharing system within this environment.
(iii) Specifying the goals of this sharing system.
(iv) Identifying the stakeholders of this sharing system.
(v) Identifying the performance measures, and instilling the necessary guarantors for the desired performance.
(vi) Identifying the enemies of the sharing system.

(vii) Mapping out all the above, and, defining how interactions between them could be reinforced with new design interventions, for instance, through the design of a new sharing culture.

References

S. Armondi, S. Di Vita, Contemporary production, innovative workplaces, and urban space: projects and policies. J. Urban Technol. **24**(3), 1–3 (2017)

B.H. Banathy, *Designing Social Systems in a Changing World* (Springer, New York, NY, 1996)

S. Boh, Two more hawker centres to get waste digesters, in *The Straits Times* (2017), https://www.straitstimes.com/singapore/environment/two-more-hawker-centres-to-get-waste-digesters

K.E. Boulding, General systems theory: the skeleton of science. Manag. Sci. **2**, 197–208 (1956)

J.P. Carse, *Finite and Infinite Games: a Vision of Life as Play and Possibility* (The Free Press, New York, NY, 1986)

J.K.H. Chan, Y. Zhang, Sharing space: urban sharing, sharing a living space, and shared social spaces. Space Cult. (2018). https://doi.org/10.1177/1206331218806160

J.K.H. Chan, The ethics of working with wicked urban waste problems: the case of Singapore's Semakau Landfill. Landsc. Urban Plan. **154**, 123–131 (2016)

Channelnewsasia, LTA rolls out bicycle parking zones. *Channelnewsasia* (2017), https://www.channelnewsasia.com/news/singapore/lta-rolls-out-bicycle-parking-zones-8599924

C.W. Churchman, *The Systems Approach* (Delta Books, New York, NY, 1968)

C.W. Churchman, *The Design of Inquiring Systems: basic Concepts of Systems and Organization* (Basic Books, New York, NY, 1971)

C.W. Churchman, *The Systems Approach and Its Enemies* (Basic Books, New York, NY, 1979)

D. Ciaffi, Sharing the commons as a 'new top' of Arnstein's ladder of participation. Built Environ. **45**(2), 162–172 (2019)

B. Cohen, P. Munoz, Sharing cities and sustainable consumption and production towards an integrated framework. J. Clean. Prod. **134**, 87–97 (2016)

M. Cox, G. Arnold, S.V. Tomás, A review of design principles for community-based natural resource management. Ecol. Soc. **15**(4), 38 (2010). https://www.ecologyandsociety.org/vol15/iss4/art38/

M. De Angelis, Grounding social revolution: Elements for a systems theory of commoning, in *Perspectives on Commoning: autonomist Principles and Practices*, ed. by G. Ruivenkamp, A. Hilton (Zed Books, London, UK, 2017), pp. 213–254

E. Durkheim, What is a social fact?, in *The Rules of Sociological Method*, ed. by S. Lukes (The Free Press, New York, NY, 1982), pp. 50–59

A. Escobar, *Designs for the Pluriverse: radical Interdependence, Autonomy, and the Makings of Worlds* (Duke University Press, Durham, NC, 2017)

B.M. Frischmann, *Infrastructure: the Social Value of Shared Resources* (Oxford University Press, Oxford, UK, 2012)

R.E. Goodin, Institutions and their design, in *The Theory of Institutional Design*, ed. by R.E. Goodin (Cambridge University Press, New York, NY, 1996), pp. 1–53

A.O. Hirschman, *Exit, Voice, and Loyalty: responses to Decline in Firms, Organizations, and States* (Harvard University Press, Cambridge, MA, 1970)

A. Huron, Working with strangers in saturated space: reclaiming and maintaining the urban commons. Antipode **47**(4), 963–979 (2015)

N.A. John, in *The Age of Sharing* (Polity, Malden, MA, 2017)

E. Katrini, Sharing culture: on definitions, values, and emergence. Sociol. Rev. Monogr. **66**(2), 425–446 (2018)

E. Katrini, in *Creating the Everyday Commons: spatial Patterns of Sharing Culture*. Unpublished Doctoral Dissertation (Carnegie Mellon University, Pittsburgh, PA, USA, 2019)

M. Kornberger, C. Borch, Introduction: urban commons, in *Urban Commons: Rethinking the City*, ed. by C. Borch, M. Kornberger (Routledge, New York, NY, 2015), pp. 1–21

A. Light, C. Miskelly, Sharing economy vs sharing cultures? designing for social, economic and environmental good. Interact. Des. Archit. J. **24**, 49–62 (2015)

E. Manzini, *Design, When Everybody Designs: an Introduction to Design for Social Innovation* (MIT Press, Cambridge, MA, 2015)

K. Marx, Capital: volume one, in *The Marx-Engels Reader*, 2nd edn., ed. by R.C. Tucker (W.W. Norton & Company, New York, NY, 1978), pp. 294–438

D.H. Meadows, *Thinking in Systems: a primer* (Chelsea Green Publishing, White River Junction, VT, 2008)

J. Mingers, *Self-producing Systems: implications and Applications of Autopoiesis* (Plenum Press, New York, NY, 1995)

E. Ostrom, *Governing the Commons: the Evolution of Institutions for Collective Action* (Cambridge University Press, New York, NY, 2006)

D. Petrescu, C. Petcou, C. Baibarac, Co-producing commons-based resilience: lessons from R-urban. Build. Res. Inf. **44**(7), 717–736 (2016)

C.M. Rose, The comedy of the commons: commerce, custom, and inherently public property. Univ. Chic. Law Rev. **53**(3), 711–781 (1986)

L. Sargisson, Swimming against the tide: collaborative housing and practices of sharing, in *Sharing Economies in Times of Crisis: practices, Politics and Possibilities*, ed. by A. Ince, S.M. Hall (Routledge, London, UK, 2018), pp. 145–149

D.A. Schön, The design process, in *Varieties of Thinking*, ed. by V.A. Howard (Routledge, New York, NY, 1990), pp. 110–141

W.R. Scott, *Organizations: rational, Natural, and Open Systems*, 3rd edn. (Prentice Hall, Englewood Cliffs, NJ, 1992)

H.A. Simon, *The Sciences of the Artificial*, 3rd edn. (MIT Press, Cambridge, MA, 1996)

B.F. Skinner, *Beyond Freedom and Dignity* (Hackett Publishing Company, Indianapolis, IN, 2002)

S. Stavrides, *Common Space: the City as Commons* (Zed Books, London, UK, 2016)

C. Teh, PUB, NEA find a way to convert sludge and food waste into energy. in *The Straits Times* (2019), https://www.straitstimes.com/singapore/pub-nea-find-a-way-to-convert-sludge-and-food-waste-into-energy

W. Ulrich, Can we secure future-responsive management through systems thinking and design? Interfaces **24**(4), 26–27 (1994)

P. Zapata, M.J.Z. Campos, Producing, appropriating and recreating the myth of the urban commons, in *Urban Commons: rethinking the City*, ed. by C. Borch, M. Kornberger (Routledge, New York, NY, 2015), pp. 92–108

Chapter 5
Sharing Ethics

On that fateful December morning, Mr. Heng Lay Peng was traveling to work on his motorcycle on the Seletar Expressway in Singapore. Soon after, he met with a fatal accident. While exact details of this accident are still pending investigation, initial reports, however, revealed that a moving trailer had ran over and crushed him, dragged his body across the expressway, and leaving a gory trail in its wake (Wong 2019). To the distress of his family members, vivid videos of this fatal and gruesome accident, captured from multiple angles by commuting motorists, soon surfaced online. Despite the urgings of the police, these video clips continued to be shared on social media and chats platforms (Wong 2019). These shared video clips were justified on the grounds that they could exhort safer driving. But others opposed sharing these clips, arguing that they could offend the deceased's dignity, and moreover, exposed Mr. Heng's family to experience his woeful death many times over—perhaps as many times as these clips have been shared.

But how can we know which position is more justifiable, or perhaps even, right? What are the stakes in each of these positions? And when is sharing wrong? Based on the report by Wong (2019), doing the right thing matters. To advocate sharing these video clips on the grounds that they play an admonishing role for safer driving means that the late Mr. Heng's family may never come to a state of closure on his demise. But to protect the dignity of the late Mr. Heng and to respect the rights of his family by removing every single video clip on the Internet may mean violating important rights to information—especially information that has been argued to possess some value for public safety. On this controversial matter of sharing, the expectation that safer driving would result through video sharing has clashed with the obligation of privacy and dignity for the deceased and his family. Such conflicts that arise from sharing are not unique. It is also readily observable in other sharing practices as well. Consider the practice of maximal sharing, or 'sweating the asset' in a sharing arrangement of Airbnb. While this practice satisfies the practical interests of the owner or other beneficiaries of this asset, this practice can, however, conflict with an obligation not to unduly burden one's neighbors, who now have to live with the inconveniences that

J. K. H. Chan and Y. Zhang, *Sharing by Design*, SpringerBriefs in Applied Sciences and Technology, https://doi.org/10.1007/978-3-030-43569-1_5

are brought about by a relentless stream of 'tenants' with very different occupancy demands and behaviors.

These conflicts that are brought about by sharing not only represent a clash of different informed opinions, but they also reflect a complex collision of different ethical positions. One conception of the ethical is taking the relation of our actions to the demands, needs, claims, desires, and lives of other people seriously (Williams 2006: 12). To be plainly unethical then is to renounce that our actions have any ethical claim on other people. Ethical conflicts in sharing are therefore alarming to the extent that they dispel the notion that sharing is an inherently cooperative behavior with others; that when further aggregated and consistently organized, also yields an inherently ethical practice that can do no harm. Furthermore, sharing has attained a positive glow in contemporary culture. For example, John (2017) notes that sharing has been recently conceived as a morally superior practice, and Hall and Ince (2018: 1) report that sharing has been positioned as inherently beneficial such that its innate goodness is rarely questioned. In this manner of imagining sharing as morally lofty, one might draw the inevitable conclusion that such a practice cannot be unethical, or at the very least, is expected to be free from the complexity and messiness of ethical conflicts.

To share is not an unequivocal good, and to hold a morally lofty perspective of sharing is not only inaccurate but also imprudent for the following two reasons. First, and as prior examples demonstrate, this lofty perspective is false; ethical conflicts do exist in sharing practices and permitting them to fester inadvertently leads to an intractable deadlock, which can corrode convivial relations between individuals. A part of our ethical life is shared with others (see Williams 2006: 191), and sharing is a specific form of social interaction that renders the complexity and the messiness of this ethical life more apparent, and where remedying action by ethical know-how is likely to become even more necessary. Even so, ethical knowledge is not the panacea for moral complexity and struggling with conflicts is an unavoidable part of contemporary social life characterized by the plurality of persons and evaluative frameworks (Kekes 2019: 287). Nevertheless, ethical knowledge offers the required tools that can clarify this complexity—especially in the company of others—that can make a flourishing ethical life together more likely. Second, the designers of the sharing system have to reckon with real complexity and messiness of actual behaviors and social interactions, and this means taking ethical conflicts seriously because they have a tendency to spiral into schisms and fractions that can undermine the sharing system. In sum, it is important for the designer to frame sharing with its confounding complexity and messiness and in tandem, to initiate a deeper engagement with ethics in the design of the sharing system.

5.1 Why Is Ethics Important for Sharing?

Beyond these conflicts, there are at least three further reasons on why ethics, or ethical knowledge, is important for sharing practices. First, as mentioned in Chap. 1, the fact that sharing has occurred neither presumes sociality and positive social relations, nor would sharing be automatically aligned to a more just redistribution of resources (Davies and Evans 2018). As a matter of fact, sharing platforms have been observed to worsen inequality—for instance, when they tend to create a risk shift from firms to workers (Schor and Attwood-Charles 2017), and where consequentially, these platform 'gig' workers are derived of a permanent connection to the labor market and working benefits, ultimately resulting in precarity and exploitation (Bates et al. 2019: 436). These inequitable dynamics also extend well beyond transactional sharing practices. Following Widlok's (2017) studies of traditional sharing practices, he argues that sharing is not reciprocal, and it tends to be characterized by an imbalance of resource transfer from net providers to net receivers. While this does not mean that sharing has to always presume an unequal distribution of resources, it does however suggest that the starting point of many sharing practices hinges on providers who share their bountiful resource with others that do not have access to these resources. After all, it is impossible to share what one does not already have. The social dynamics of this imbalance between net providers and net receivers in sharing then point to the need for sharing to be enacted as a form of "moral skill rather than a technical capacity"—a sense of knowing the appropriate thing to do in a particular situation (Widlok 2017: 111). For the resource provider, this moral skill would then provide the necessary tact amid evident inequality, combined with the savvy to maintain the dignity of the receivers as they receive their allotment. But what exactly is this moral skill? Is it comprised of virtues, principles, or what Varela (1999: 24) refers to as a form of "ethical expertise"?

Second, sharing can paradoxically increase risk-taking behaviors. This paradox can be observed in the following two instances. In the first instance, sharing between net providers and net receivers is characterized by some degree of the principal-agent problem, where these providers are unable to monitor or control all aspects of the receivers' behaviors, and where this monitoring gap allows certain receivers to behave opportunistically in ways that tend to promote their interest at the expense of the providers' (see de Graaf and Wiertz 2019: 42). Consequentially, this can result in a situation of moral hazard where certain receivers now have an incentive to engage in risky behavior while the providers then end up bearing the cost of this risky behavior (de Graaf and Wiertz 2019: 44). Corroborating this reading, an empirical study on the motivations of participants in the sharing economy reveals a difference between the motivations of receivers and providers (Böcker and Meelen 2017). While the receivers of shared resources tend to be driven by economic motivations, the motivations of the providers of these resources are more mixed and include altruistic and community-oriented elements (Böcker and Meelen 2017: 37). If such economic motivations tend toward the maximization of an altruistically provided resource—akin to something like resource abuse—by certain receivers, this will surely come at the expense of the providers that now have to bear the cost of maintaining a highly exploited resource in order to sustain an ongoing sharing practice.

In the second instance, individuals can engage in risk-pooling by sharing their resources, which in turn paradoxically permits them to take on riskier but more rewarding ventures. Researchers discovered that when faced with variable catches and other uncertainties, fishermen in Toyama Bay, Japan, proceeded to share income, information, and training (Bowles and Gintis 2005: 383). By combining their resources and by sharing both incomes and costs of their livelihood, these fishermen were able to fish in much riskier but higher yield locations. In this case, risk-pooling was certainly motivated by the incentive of greater profitability for everyone; but at the same time, risk-pooling also provides the insurance, or securitization, for riskier ventures in higher yield locations. To formulate this slightly differently, the capacity to share risks has paradoxically encouraged greater risk-taking behaviors. In sum, sharing practices can increase risk-taking behaviors on the part of sharing participants, and this ought to be accompanied by a commensurate ethical oversight, which is however mostly absent in sharing discourses.

Third, as discussed in Chaps. 3 and 4, social formations and spatial configurations that reinforce or perpetuate sharing behaviors and relations can be designed. But this ability to design a sharing system comes with the obligation to justify why such a system ought to be prioritized. In other words, to materialize sharing by design is no different than promoting sharing behaviors and normalizing sharing relations, which are normative propositions. These normative propositions have to be justified. And given that sharing by design always reinforces certain social possibilities while foreclosing others, it is paramount that its design specifications are given consent by plural sharing participants. To clarify and to facilitate the process of justifying certain design specifications over others, it is important to weigh claims and counterclaims through careful ethical analyses. Therefore, in at least these three ways, ethics is an indispensable knowledge for the designer of the sharing system.

5.2 Sharing Ethics: A Preliminary Outline

Obviously, 'sharing ethics' can refer to the platitude that participants of a sharing system share a common body of ethical codes, values and principles. In this reading, 'sharing ethics' is merely a shorthand phrase for individuals who share a similar set of values, virtues, or principles in the same sharing system. But this is not the only notion of ethics in this chapter. Instead, 'sharing ethics' refers to the broader body of ethical knowledge pertaining to the human activity and practice of sharing. It is also this ethical knowledge that the designer can use to design a sharing system, where it can also be applied to sustain and regulate newly designed sharing systems.

In highlighting this broader approach to sharing ethics, this chapter will examine sharing through the moral philosophical categories of virtues, principles, and ethical expertise. The primary question that guides this chapter is: What are the categories of ethical knowledge that one can expect to encounter in sharing activities, practices, and systems? The short answer, which will be further elaborated in subsequent sections, consists of the following three categories: in any sharing practice, one is likely to encounter virtues of sharing, principles of sharing, and ethical expertise involved in the act of sharing. Ineluctably, there is also a fourth category, which due to the limited scope of this chapter can only be described briefly here. This fourth category entails the hard problems of intractable dilemmas: What if there are dilemmas of sharing that are neither amenable to virtuous actions nor fair principles, and are furthermore beyond the recourse of ethical expertise? In other words, what if sharing entails tragic moral costs? But we aver that these genuine moral dilemmas are rare and will therefore focus our discussions instead on the more general categories of virtues, principles, and ethical expertise—all that are likely to be encountered in the design of sharing systems. The sum of these discussions then constitutes the preliminary outline for the important, but underexplored area of sharing ethics.

5.3 Three Key Distinctions

Despite the relevance of ethics in sharing activities and practices, the ethical dimension of sharing behaviors and relations remains an underdeveloped knowledge area. To facilitate a further discussion on sharing ethics, it is important here to lay out the following three key distinctions as a primer for subsequent discussions.

First, it is important to make a distinction between an internal ethics that facilitates sharing behaviors and relations within the sharing system, and an external ethics that regulates this sharing system in relation to other sharing systems and its own environment (Chan 2019: 154). While internal ethics suggests the virtues, principles, and moral skills that can motivate, regulate, and streamline sharing behaviors within the system, external ethics aims to evaluate the impacts and consequences of the sharing system on other sharing systems and its own environment. Clearly, the prevailing internal ethics—especially rules and principles that are used to regulate the sharing system—will influence what counts as permissible or impermissible impacts and consequences on other sharing systems or its own environment. Internal ethics and external ethics are therefore overlapping categories. This said, for the purpose of design, their differences matter more than their intersections, and it is prudent to make this important distinction. Here, it is possible to suggest that external ethics of sharing systems tends to revolve around two primary focuses. The first focus centers on the evaluation of impacts and consequences imposed by a sharing system on its environment, and the second focus emphasizes on clarifying the dilemmatic conflict with another equally valuable or legitimate sharing system (see Harvey 2012: 70)—where only one could flourish. This chapter, however, will focus on the internal ethics of the sharing system because this is the area that comes under the immediate control of the designer.

Second, and following from the idea of internal ethics, it is possible to distinguish between autotelic and telic sharing. Autotelic sharing entails sharing as an act for its own sake (Widlok 2004: 61), while telic sharing revolves around instrumentalizing the act of sharing for some other goals. For Widlok (2004: 61), an example of autotelic sharing is sharing food with neighbors, relatives, or anyone who happens to be around for the sake of shared enjoyment from what is being shared. To describe this in a clear but still roundabout way, autotelic sharing is sharing that constitutes its own purpose. On the other hand, telic sharing is defined as sharing as a means to some other ends. For instance, an individual could share certain resources with others with the goal of alleviating their perceivable needs or perhaps, to boost her own ego. This form of sharing is usually not mistaken for the joy of autotelic sharing for its own sake. However, this is not to suggest that in actual sharing practices, autotelic and telic sharing acts are always mutually exclusive. Sharing in real life is usually driven by a mix of different motivations (Böcker and Meelen 2017), and there is no rule to stipulate that one cannot enjoy the act of sharing while alleviating the material needs of others. However, similar to the first distinction, drawing this second distinction is important because the ethical considerations of autotelic sharing tend to differ from that of telic sharing. For instance, telic sharing often requires ethical justification: to find out if the goals of sharing are justifiable, or if the means employed for these goals can be justified. Therefore, for telic sharing, it is important to answer these questions: To what extent is sharing justified? Do the ends justify the means of sharing—for instance, does extending the interest of public safety justify sharing the video clips of the gruesome traffic accident? In contrast, we are not likely to apply a similar ethical evaluation to autotelic sharing. Instead, we may try to describe, or appraise, the moral worth in autotelic sharing. We may, for instance, try to find out if autotelic sharing has substantively elevated the quality of joy or trust in the community.

Third, and following from the second distinction, one could further distinguish between the moral ideals of sharing, and the ethics of sharing. For better or worse, the enterprise of the sharing economy has been nearly driven by assuming the former. On this, Slee (2015: 16) argues that the sharing economy has appealed to ideals such as equality, sustainability, and community, but has mostly resulted in the opposite—where in the name of sharing, massive private fortunes have been amassed, more precarious forms of labor have materialized, and real communities corroded. On the other hand, the ethics of sharing entails the ethical knowledge, or else the deliberative process, that tries to justify these moral ideals especially when they clash. In other words, the ethics of sharing concerns the application of ethical knowledge to the careful deliberation of moral matters in sharing activities and practices. For example, different participants of a sharing system are likely to be motivated by different moral ideals of sharing. Some might be driven by what they perceive to be egalitarianism in sharing practices, while others are motivated by the more sustainable ways of life promised by a sharing culture. But these ideals can clash in practice. What if a more sustainable sharing culture implies a compromise on egalitarianism, where a minority of individuals who already possess more must make disproportionate sacrifices so that they then receive a lesser share compared to others? It is in such a situation that the ethics of sharing can intervene to clarify, inform, and perhaps also, to offer a way forward.

5.4 Sharing Ethics as Virtues

In recognizing that sharing is often a non-reciprocal, one-way resource transfer (Widlok 2017), one might also be led to wonder what motivates this type of resource transfer. Empirical research suggests that while motivations are often mixed, there is nevertheless some elements of altruism involved (Böcker and Meelen 2017: 37). If so, then one could ask: what may be a virtue that must exist in order for altruism—defined simply here as unconditional giving without the expectation of reciprocity—to be possible? Across history, this virtue has adopted many names. Schopenhauer (2010: 228) refers to it as "*caritas*", or the "virtue of loving kindness". More contemporarily, it has been defined as the virtue of benevolence (Frankena 1973: 64). And underlying benevolence is compassion, where the agent aims to help as many people as much as he or she can, and where this same agent perceives another person's suffering as the immediate motive of his or her remedying action (Schopenhauer 2010: 229). While virtues in practice inevitably involve actions, it is nevertheless important to emphasize that virtues are about dispositions and traits—virtues are about being a certain kind of person (Frankena 1973: 64). For this reason, the virtue of benevolence is distinguished from the principle of beneficence, which entails an obligation, or a requirement, to do good (Frankena 1973). Applying this to sharing then, an individual characterized by the virtue of benevolence is predisposed to be compassionate, and will perhaps even prioritize the needs of others ahead of her own. Because virtues are not rules, they do not exactly specify what the individual ought to do. This benevolent individual will sometimes share more, and sometimes less; but by benevolence, this individual would almost certainly share—and often at her own expense.

Construed in these ways, virtues are highly significant qualities in sharing by design. Recalling the discussion on system performance in Chap. 4, clients—or the people that the sharing system serves—are also the bearers and standards of performance measures of the sharing system. Following this, if individuals embody the virtue of benevolence in this system, then this sharing system is likely to perform according to the wonderful capabilities and possibilities that benevolence can afford. Even so, for a sharing system is to be sustained over time, it is important that resource exchanges are also balanced over time. In other words, a sustained sharing system cannot depend wholly on the presence and the readiness of benevolent providers alone. After all, reciprocal exchanges are often needed to sustain a particular practice or institution (Becker 1986: 106). While altruistic providers need not be reciprocated in kind, they are surely boosted by receivers of the shared resource stepping up, from time to time, to help others. This aspect of reciprocity is deemed especially important in newly configured sharing systems, where both providers and receivers are strangers to each other, and where trust and social capital are low. If compassion and benevolence are reciprocated by being expressed in different ways within the sharing system, then this sharing system is likely to have a better chance of flourishing.

On this note, Becker (1986: 93) thinks that reciprocity is a virtue, but a recipient's virtue. However, the virtue of reciprocity presents a special problem in sharing ethics. Consider two prototypical situations commonly found in sharing practices. First, there are moments when a virtuous recipient is unable to reciprocate—for instance, when a provider makes an anonymous resource donation to the sharing community and then disappears. In this situation, the virtuous response is to emulate the benevolence that the recipient has experienced, and in this process, consolidate a pattern of social life within the sharing community where benevolence can find different outlets of expression (see Becker 1986: 93–94). In other words, the virtuous recipient, who is unable to reciprocate in kind, practices benevolence for others as an expression of gratefulness, which may consolidate into a new pattern of social life within the sharing the community. Second, in another situation, a sharing system is often jointly constituted by different individuals working together. Some may contribute more, and others less. In this "joint project", how should one reciprocate for the disproportionate benefits that one has received or for the disproportionate sacrifices made to produce these benefits (Becker 1986: 113)? On this point, Widlok (2004: 62) is surely correct that these likely situations found in sharing practices do not, and cannot, lend itself to the creation of obligations for future transfers. But if the recipients or beneficiaries of sharing are predisposed to reciprocate for the benefits that they have received, then reciprocity may not be defined in terms of obligations, or what one is required to do, but more in line with what one wants to do. In this way, while sharing appears as a non-reciprocal and one-way transfer of resources, it may also be creating the potential for volitional and innovative acts of reciprocation to emerge if Becker is also correct.

5.5 Sharing Ethics as Principles of Justice and Beneficence

According to John (2017), rules are important especially when sharing is seen as a form of distribution and division of resources. What is considered a fair share? And how can fairness in the distribution and the division of resources be ensured? Ineluctably, sharing is conceptualized as an ethical practice because of this relation to the fair distribution of resources (John 2017: 6–7). In devising the effective rules that can ensure fairness, the designer of the sharing system would have to deduce them from more foundational principles. It is at this level that the designer would encounter sharing ethics in the form of principles of justice and beneficence. Unlike virtues, principles entail duties or what one ought to do (Frankena 1973: 64). Following this, a general principle of justice may stipulate, 'one ought to treat everyone equally', while a general principle of beneficence may state, 'one ought to do good and never do harm'. These general principles are merely illustrative; complex sharing systems are likely to require more complex principles. For a start, it is important to state that different types of sharing require different types of rules, which in turn, may mean invoking different principles. We illustrate this point here by briefly reiterating two main types of sharing (see also Chap. 4).

The first type of sharing is zero-sum sharing, which entails sharing as a way to divide and distribute tangible and limited goods. In zero-sum sharing, the total value of sharing is capped: the more one shares, the less one has (John 2017). Sharing the same tangible and divisible resource more means each recipient receiving an even smaller share. And for tangible but non-divisible goods, if one is using the vacuum-cleaner from the tool-library, then another person is deprived of using it. In these ways, zero-sum sharing is not only about fairness of distribution (i.e. what is one's fair share?) but also about fair competition (i.e. who gets what and is the competition fair?). On the other hand, the second type of sharing is non-zero-sum sharing, where the total value of sharing is not fixed. Non-zero-sum sharing usually entails the sharing of non-subtractable, non-excludable, and indivisible goods, such as knowledge, news, information, and sentiments (Chan and Zhang 2018). Sharing these goods does not diminish them but instead, can increase their values. As a matter of fact, open access to such resources can lead to scale returns (Frischmann 2012: 6). For example, allowing open access to a seminal research paper is likely to multiply readership, which may in turn generate new research trajectories and teaching activities. But it also appears that a prudent choice of open sharing is significant here. For instance, open, repeated, and relentless sharing of the video clips of the gruesome accident might have generated some social value, but it has also led to hurt feelings and public backlash.

Any sufficiently complex sharing system is likely to comprise of both types of sharing. It is, therefore, important first to consider the rules, and then the principles, that undergird these sharing types. As discussed, fair rules of division and distribution, as well as rules of fair competition, are important for zero-sum sharing. This may mean a rule where every recipient receives an equal share of a limited resource, and a rule on taking turns based on a first-come-first-serve basis for sharing a vacuum-cleaner. Conversely, for non-zero-sum sharing, a functional rule may entail imposing limits and conditions on sharing, for instance, where sharing of the video clips is only permitted within certain institutional channels, or under the stringent condition of redacting the gory details. Assuming these are examples of possible rules, then what are the principles that must exist such that these rules could be deduced?

Corresponding to these two types of sharing are namely, the principle of justice and the principle of beneficence. Specifically, for the principle of justice, it is primarily the principle of distributive justice, and entails specifying the comparative treatment of individuals (Frankena 1973: 49). If, in one instance, this comparative treatment is specified toward absolute equality by treating people equally, then the principle could be simply stated as, 'Everyone is entitled to the same share', and a corresponding rule of sharing may entail, 'Divide the resource into equal shares'. But if, in another instance, this comparative treatment is tilted toward enhancing the welfare of the least advantaged, then the principle may approach Rawls's (1971: 65) principle as, 'unequal shares based on contingent distribution are acceptable if and only if they improve the welfare of the least advantaged member in the sharing system'. On the other hand, the principle of beneficence broadly requires one to do good and not evil (Frankena 1973: 45). This does not mean an individual, who accepts this principle, must be beneficent all the time and on all occasions (Driver 2007: 92), if only because this is likely an impossible requirement. But this principle does require the individual, who for instance is about to share the video clips of the gruesome accident, to carefully think about the good that sharing these clips can do, and also the foreseeable but unintended evil that may come to pass when these clips are shared. In sum, and in contrast to rules, principles are more general but no less specific in their formulation. The designer of the sharing system is likely to find it necessary to first formulate principles for sharing, and subsequently, to deduce the relevant rules that can support the operation of the sharing system. In other words, principles are the ethical anchors of sharing systems, and unlike the virtues of the sharing community, principles can be nearly and completely specified by the designer.

5.6 Sharing Ethics as Ethical Expertise

Earlier, it was mentioned that sharing is akin to a moral skill rather than a technical capacity (Widlok 2017: 111). An individual enabled by this moral skill is able to know what to do in a particular situation because she has been immersed in her community of practice by first participating in this practice, and subsequently, imitating the experts found there (Widlok 2017: 111). This moral skill can be observed in everyday life, where people rarely have trouble helping others, responding to the needs of others, or knowing how to share. Using this moral skill is often unpremeditated, and Fisher (2019: 83) offers an account here:

> Helping others doesn't have to be planned ahead; plenty of opportunities exist right in front of us. I walked outside my office the other day to see a seriously inebriated man attempting to get on his bicycle and nearly falling into a street with passing cars, and not long after that, walking away from a meeting in a local coffee shop, I saw a man in an electrified wheelchair tipped over in a parking lot, with the food in the bag he had carried spilled out across the pavement. In both cases, I had other places to go, but I could not leave them that way…
>
> I called the police for help with the drunk and stayed with him until they arrived so that he wouldn't get hit by a car or fall over on his bike, while I and a colleague of mine helped the disabled man back up in his chair and gathered his food and got him on his way…

Varela (1999: 23–24) would refer to this ability to immediately cope with the ethical demand of the situation as a form of "ethical expertise". In the situations described, Fisher did not stand back from the scene to reason if he would help or to decide how exactly to help. Instead, he responded in a manner that would strike anyone as commonsensical: to call for more help; stay with a vulnerable person; help a fallen man up; send him on his way. But what this commonsensical description does not say is the necessary tact, savvy, or *savoir faire* involved: Should one call for the police immediately or try to establish some rapport first? And how should one help an impaired man up to his wheelchair without afflicting his dignity? Fisher did not elaborate. But any successful attempt to help or to respond to someone's needs has to presume some degree of ethical expertise, which is a skilled capacity to exercise a conscientious consideration for others (Varela 1999: 35). In this way, ethical expertise begins to explain how people, in ordinary situations, know exactly what to share, and how to share, without much trouble.

The only trouble, however, is extreme situations that threaten this ethical expertise. Consider Hardin's (1974) famous lifeboat thought experiment. To summarize this thought experiment, imagine that there are 50 individuals already on a lifeboat with a maximum capacity for 60 persons. This lifeboat is adrift at sea with more than 100 people still in the water. The two hard problems are therefore these: First, who, among the people still in the water, should be allowed to share the precious space on the lifeboat up to its maximum capacity? And second, on what grounds should the rest be rejected from the lifeboat? As Hardin (1974) argues, if everyone is allowed into the lifeboat, then the lifeboat sinks and everyone dies. Clearly, these are hard ethical problems that threaten the functioning of ethical expertise. Varela (1999) did not reveal how ethical expertise might cope with these extreme situations characterized by hard moral dilemmas. Following philosopher Martin Heidegger (1889–1976), Winograd and Flores (1987: 36–37) characterized such extreme situations as a form of breakdown, where the individual can no longer cope by relying on her ethical expertise, but must now stand back from action in order to reflect on the ethics of the available options. But no matter the choice, this 'lifeboat thought experiment' suggests an inevitable moral loss.

Nevertheless, it is reasonable to suppose that such extreme situations are rare in ordinary life where sharing systems are positioned to exist. For the designer of the sharing system, the idea that ethical expertise is present in sharing may then suggest the following two ideas. First, the designer may wish to design configurations or spaces where participants can repeatedly gather to interact and learn from each other—a concrete way of actualizing a community of sharing practice where there are teachers and students of the ethical expertise of sharing. And second, the designer may need to carefully scrutinize the sharing system to ensure that such extreme situations like the hypothetical 'lifeboat thought experiment' do not emerge in everyday operations.

References

L.K. Bates, A. Zwick, Z. Spicer, T. Kerzhner, A.J. Kim, A. Baber, J.W. Green, D.T. Moulden, (eds.), Gigs, side hustles, freelance: what work means in the Platform Economy/Blight or remedy: Understanding Ridehailing's role in the precarious "Gig Economy"/Labor, gender and making rent with Airbnb/The gentrification of 'sharing': From bandit cab to ride share tech/The 'sharing economy'? Precarious labor in neoliberal cities/where is economic development in the platform city?/shared economy: WeWork or we work together. Plan. Theory Pract. **20**(3), 423–446 (2019)

L.C. Becker, *Reciprocity* (Routledge, New York, NY, 1986)

L. Böcker, T. Meelen, Sharing for people, planet or profit? analysing motivations for intended sharing economy participation. Environ. Innov. Soc. Trans. **23**, 28–39 (2017)

S. Bowles, H. Gintis, Social capital, moral sentiments, and community governance, in *Moral Sentiments and Material Interests: the Foundations of Cooperation in Economic Life*, ed. by H. Gintis, S. Bowles, R. Boyd, E. Fehr (MIT Press, Cambridge, MA, 2005), pp. 379–398

J.K.H. Chan, Y. Zhang (2018). Sharing space: Urban sharing, sharing a living space, and shared social spaces. Space Cult. https://doi.org/10.1177/1206331218806160

J.K.H. Chan, *Urban Ethics in the Anthropocene* (Palgrave Macmillan, Singapore, 2019)

A. Davies, D. Evans, Urban food sharing: emerging geographies of production, consumption and exchange. Geoforum (2018). https://doi.org/10.1016/j.geoforum.2018.11.015
N.D. de Graaf, D. Wiertz, *Societal Problems as Public Bads* (Routledge, New York, NY, 2019)
J. Driver, *Ethics: the Fundamentals* (Blackwell, Malden, MA, 2007)
T. Fisher, *The Architecture of Ethics* (Routledge, New York, NY, 2019)
W.K. Frankena, *Ethics*, 2nd edn. (Prentice Hall, Upper Saddle River, NJ, 1973)
B.M. Frischmann, *Infrastructure: the Social Value of Shared Resources* (Oxford University Press, Oxford, UK, 2012)
S.M. Hall, A. Ince, Introduction: Sharing economies in times of crisis, in *Sharing Economies in Times of Crisis: practices, Politics and Possibilities*, ed. by A. Ince, S.M. Hall (Routledge, New York, NY, 2018), pp. 1–15
G. Hardin, Lifeboat ethics: the case against helping the poor (1974), https://www.garretthardinsociety.org/articles/art_lifeboat_ethics_case_against_helping_poor.html
D. Harvey, *Rebel Cities: from the Right to the City to the Urban Revolution* (Verso, London, UK, 2012)
N.A. John, *The Age of Sharing* (Polity, Malden, MA, 2017)
J. Kekes, *Hard Questions: facing the Problems of Life* (Oxford University Press, New York, NY, 2019)
J. Rawls, *A Theory of Justice* (Harvard University Press, Cambridge, MA, 1971)
A. Schopenhauer, *The Two Fundamental Problems of Ethics* (Oxford University Press, New York, NY, 2010)
J.B. Schor, W. Attwood-Charles, The "sharing" economy: Labor, inequality, and social connection on for-profit platforms. Sociol Compass (2017). https://doi.org/10.1111/soc4.12493
T. Slee, *What's Yours is Mine: against the Sharing Economy* (OR Books, New York, NY, 2015)
F.J. Varela, *Ethical Know-How: action, Wisdom, and Cognition* (Stanford University Press, Stanford, CA, 1999)
T. Widlok, Sharing by default? outline of an anthropology of virtue. Anthropol. Theory **4**(1), 53–70 (2004)
T. Widlok, *Anthropology and the Economy of Sharing* (Routledge, New York, NY, 2017)
B. Williams, *Ethics and the Limits of Philosophy* (Routledge, New York, NY, 2006)
T. Winograd, F. Flores, *Understanding Computers and Cognition: a New Foundation for Design* (Addison-Wesley, Boston, MA, 1987)
C. Wong, Stop sharing videos of crash, experts urge. *The Straits Times* (2019), https://www.straitstimes.com/singapore/stop-sharing-videos-of-crash-experts-urge

Chapter 6
Teaching Sharing by Design

Teachers cannot avoid the responsibility for guiding the learning process. However, knowing the subject area and teaching it effectively are quite separate matters (Cahn 2018). Both of us are designers and also teachers. We think that any system of new knowledge is incomplete unless ideas about how to teach this knowledge are also included. It is apparent that from the previous three chapters, sharing by design ought to be considered as a new approach to envision, conceive, and implement more effective, efficient, aesthetically pleasing, and ethical sharing configurations. Like any new knowledge area, the question of how sharing by design should be taught is critical for its present viability and future evolution. If sharing is perceived as a practice that can effectively contribute to overall sustainability (Balaram 2018), but if the transition to a society characterized by systemic sharing is a process that will take time, then present and future generations of students should be considered in the overall sharing system and this process. Students are the guarantors that can ensure the continuity of this impactful knowledge. Framed in this way, the pedagogy of sharing by design becomes an integral component in any complex sharing system. For this reason, we aim to answer the following key question in this chapter: How should we teach sharing by design?

Specifically, we answer this question by first revisiting the theoretical outlines of the systems approach (see Churchman 1968), and subsequently, by reflecting on how these theories have been applied to conducting a postgraduate urban design studio of the Master of Architecture program at the Department of Architecture, National University of Singapore.

J. K. H. Chan and Y. Zhang, *Sharing by Design*, SpringerBriefs in Applied Sciences and Technology, https://doi.org/10.1007/978-3-030-43569-1_6

6.1 The Architectural Studio as the Site of Teaching Sharing by Design

The architectural studio is the *sine qua non* of design education (Cuff 1991: 43–44). In the studio, students—individually or in small groups—meet their professors at least twice a week to discuss the development of their design ideas. And over the course of a semester, and under the supervision of their professors, students would propose solutions to their independently framed design problems, usually beginning by framing the problem systematically, then analyzing and evaluating patterns of interactions in these problems, before settling on the development of a conceptual architectural idea. It is possible to describe the architectural studio pedagogy as a relentless spiral of problem-finding and problem-solving, where students successively discover and solve more advanced problems as they delve deeper into their design projects. On this, the architectural studio offers a direct one-to-one interaction between faculty and students that is found in few other disciplines, and it has remained the quintessential component in all architectural design education (Koch et al. 2006).

Even so, there are urgent calls to reform the pedagogy of the architectural studio. One particularly trenchant critique suggests that this pedagogy is mostly about the cultivation of cultural capital and disposition—that is, where students inherit the cultural dispositions of their professors on what they consider tasteful or distasteful in architectural design (Stevens 1998). Another critique suggests the inscrutability of the design process, where conversations between the professor and the student can turn 'elliptical' and opaque to outsiders as the design becomes more complex (Schön 1983: 81)—ironically just when complexity calls for more open and communicable dialogs. Further critiques of this pedagogy argue that the architectural studio can lack a clearly defined design methodology, where students focus on architectural form making at the expense of ignoring the systematic design process altogether (Bashier 2014). These critiques can be summed up as a pressing mandate to rethink what, and how, to teach the architectural studio.

In tandem, designers are also increasingly engaged in responding to broader societal problems, and there is therefore a need for design education to reflect this change (Scupelli 2019: 112). These societal problems are distinguished from mere social problems by the fact that a substantial group of citizens perceives a certain state of affairs as problematic (de Graaf and Wiertz 2019: 1). Examples of such societal problems are social inequality, environmental pollution, public health, and climate change, among others. These societal problems are often 'wicked problems'—'system' problems that exist on multiple scales, where they are interconnected and interdependent, and furthermore, are continually evolving and cannot be solved by any unitary solution from a single expert, discipline, or profession (Irwin 2019: 150). Importantly, wicked problems are characterized by the contestations of multiple politics and plural values (Rittel and Webber 1973), and therefore requires collaboration between different professional expertise and also expansive community engagement. For the architectural studio to remain socially relevant, it will require nothing less than changing the theory and practice of this pedagogy in order to address such societal problems.

These critical assessments of the architectural studio form the background of our attempt to teach sharing by design. On this, we argue that an architectural studio centered on sharing by design offers new opportunities to reform the studio pedagogy—especially in necessitating the systems approach. Not only does such a studio focus on the study of certain societal problems ('system' problems), but this studio would also require systematically dissecting the different system components and their interactions at any one scale. In other words, a successful scheme by the student would offer a convincing system-level solution for the selected societal problem, and moreover, this solution would entail a functionally effective and aesthetically pleasing architectural form that fulfills the disciplinary demands of architecture.

Even so, our attempt is characterized by one dilemmatic limitation often found in all studio pedagogies centered on actual communities: should the studio aim for hypothetical solutions, mostly devoid of actual interactions with stakeholders (who often cannot come together for this educational purpose because of practical reasons) and real impacts, or should the studio aim to build relationships with an actual community without knowing exactly where this would lead to, and what design problems to focus on at the outset (see Rohrbach and Steenson 2019: 260)? In other words, this is about making a choice between a design solipsism that is, however, not without the certain merits of control and closure, and the uncertainty of engaging in the real world characterized by volatile politics and incessant changes that educators have little control. In the end, we opted for the former instead of the latter, where we could teach the systems approach within a complete process of framing and then solving a sharing system problem by design.

6.2 The Systems Approach for Sharing by Design: A Framework

The systems approach was developed in recognition that complex social systems have become "far too complicated for our intellectual powers and technological capabilities to be able to really identify the central problem and determine how it should be solved…" (Churchman 1968: xi). In other words, the systems approach aims to offer some traction in asking the right questions and then making informed decisions on the design of a complex social system. The design of complex sharing systems therefore would benefit from the use of the systems approach.

Here, it is important to begin from Churchman's (1968) critiques on conventional planning and management. Table 6.1 summarizes Churchman's primary critiques of the five basic elements (or 'considerations' in Churchman's terms) in scientific planning and management, as well as our own derived implications corresponding to these critiques for the design of sharing system. We will briefly elaborate on each of these critiques before presenting a framework for the design of the sharing system.

Table 6.1 Summary of Churchman's (1968) critiques of five considerations of *scientific* systems approach and its implications to the design of the sharing system

Considerations	Churchman's critique	Implications for the design of the sharing system
Objectives of the overall system	Stated objectives are often independent of the performance	Continuous review of (real) objectives; precise measure of performance
Components (with clear and measurable sub-objectives)	Separation and rigidity that constrain communication	Develop components whose objectives are integral to those of the overall system
Environment	Imposition of environment boundaries leads to poor performance	Open environment; examining what subsumed under the environment at different scales
Resources (including money, people, time, and equipment)	Lack of adequate thinking of real resources, e.g. personal capability of personnel	Examine counter-forces and 'enemies' related to human values; develop strategies to better utilize resources
Management (considerations of the above four dimensions)	Uncertainty in receiving information about system's performance and errors to initiate changes	Dynamic view of the process of the system; effective measures to monitor the performance

Source Authors

First, Churchman's critique repeatedly highlights the gap between anticipated objective(s) and actual performance of the system, which can be affected by different elements at different stages. It means that to bridge this gap and to achieve as much of the planned objective(s) as possible, effective performance measures that can provide prompt feedback to the management of the system should be clearly identified and formulated at the onset to guide design.

Second, Churchman points out that although the rational division of the system into different task-oriented components may appear constraining or even rigid, this is nevertheless an effective way to understand and improve the operation of the system. To address this drawback of rigidity, the challenge is to establish effective communications between different components so that they do not turn into sequestered silos. On this, Churchman suggests considering the connections between the total goal of the system and the objectives of its different components. In other words, when conceiving a system, instead of directly focusing on its structure and constituent parts, we should begin by prioritizing thinking about how the total goal of the system may be dissected into more specific and achievable objectives, and then asking what kind of components can be developed and coordinated to attain these objectives and then the total goal of the system.

Third, for Churchman, the environment is not simply defined as the outside of the system. Instead, it is what "makes up the things and people that are 'fixed' or 'given' from the system's point of view" (Churchman 1968: 35). This not only suggests that there exists no clear-cut boundary between a system and its environment but also indicates that both a system and its larger environment are a part of an even larger system. Therefore, with respect to design, this means that the different components within a system are also parts of an even larger and open environment. In turn, this suggests that any opportunities and imposing constraints should be examined across multiple scales.

Fourth, Churchman's main critique of the resources, which are "the means that the system uses to do its job" (Churchman 1968: 37), is the gap between the listed resources when a system is planned, and actual resources that can be utilized in reality. This gap can be mostly attributed to human values and behaviors. This suggests that deliberate efforts should be made in the design of a system to anticipate every possible issue arising from politics, morality, religion, and conflicts that may act as counter-forces, or as 'enemies', which can hinder the performance of the sharing system by either under-consuming or over-consuming planned resources. In tandem, it is important to develop corresponding strategies to make better use of resources in order to narrow the gap between planning and implementation.

Last, the management of the system mainly deals with setting total goal and sub-objectives, developing components, allocating resources, and controlling the performance. This is a continuous process of design that requires repeated evaluation and feedback. Churchman suggests that the key to management is to obtain accurate and prompt information about the performance of the system. This suggestion echoes his first critique pertaining to the objectives of the system, calling for developing effective performance measures. For the design of the sharing system, a dynamic view of the processes of this system, as well as its management component, underpins all design considerations.

In short, by the systems approach, Churchman argues for the need to stand back and to consider the system from the broadest possible point of view. It requires continuous efforts on reviewing the performance of the system and thinking about alternatives for making informed decisions in the planning and management of the system. Based on the above analysis, we suggest that the following six questions to guide the design of the sharing system:

1. What is the sharing system? What are the total goals of this system?
2. What are the sub-systems of this system and their respective objectives? How justifiable are they? And what are the relations between the objectives of sub-systems and the total goals of the system?
3. What is the larger environment (related elements and other systems) in which this system is situated? What are the constraints and potentials from the environment?
4. What influences does this system have on the larger environment? Does this system threaten or benefit other related systems?
5. Who, or what, are the 'enemies' of this system?
6. What are the criteria for evaluating the performance this system?

6.3 Case Study: An Urban Design Studio on Sharing Cities

In this section, we apply the above framework derived from the systems approach to the pedagogical process of a postgraduate urban design studio of the Master of Architecture program at the Department of Architecture, National University of Singapore. This design studio focused on sharing neighborhoods and was comprised of eight schemes, each conceived as a sharing system inter-related with one another. In other words, students responsible for each of these schemes had to collaborate with other students responsible for a different scheme. In this way, sharing was not only the design focus; sharing was also the *modus operandi* of the pedagogy. However, to clearly demonstrate the pedagogical process, we will only focus on two of these eight schemes, instead of presenting all eight of them. We will first outline the background of this design studio and then discuss these two schemes rendered by students in this design studio in greater details.

6.4 Background of the Urban Design Studio

This design studio attempted to explore whether a sharing system is a plausible alternative to both the state-led and market-driven approaches to the regeneration of historical urban quarters. The key question underpinning this design inquiry is what kind of sharing system can be created and produced using local resources, in order to transform a historical neighborhood into a socially convivial and environmentally sustainable community.

As shown in Fig. 6.1, the site of this urban design studio is Joo Chiat, a mix-used historical urban area located close to the east coast of Singapore. An interesting feature of the site is that a number of well-known restaurants and cafés are concentrated along the north–south Joo Chiat Street located in the center of the site, and they attract people from all over Singapore throughout the year. Nevertheless, this does not suggest that the area is economically vibrant and socially convivial. On the contrary, Joo Chiat was once a hotbed for crimes and anti-social behaviors, and is now a non-descript neighborhood filled with a number of underutilized historical buildings, neglected amenities, and pocket public spaces all challenged by increasing gentrification. In other words, the neighborhood is facing the challenge of social cohesion, and this provides an ideal test bed for exploring what a sharing system can contribute.

Fig. 6.1 Joo Chiat, Singapore (*Source* Reproduced from Google Earth)

6.5 Scheme One: Sharing Infrastructure

In the first scheme, the proposed sharing system is a new infrastructure system, whereby food wastes from the entire neighborhood can be recycled and used to produce clean energy to power shared mobility and to activate public spaces. Besides improving energy efficiency and mobility of the neighborhood, the total goal of this system also includes enhancing residents' sense of belonging and fostering community participation by engaging them in both the production and operation of this sharing system. Design investigation is closely guided by this goal.

Specifically, following the systems approach, four sub-systems are proposed, of which the objectives are inter-related and also integrated to the total goal. First, a fleet of electric self-driving cars is introduced to provide shared rides for both residents and visitors of the neighborhood. Second, a waste recycling system is conceived that can collect and pre-process food wastes from across the entire neighborhood, and transport them to a centralized digester for electricity generation. This centralized digester together with the high performance batteries then constitute the energy supply system that powers the entire fleet of self-driving cars and personal mobility devices (e.g. e-scooter and Segway) as well as public facilities (e.g. basketball court and digital libraries) that are distributed in many pocket public spaces. Last and arguably the most important is a credit system that allows individuals to cumulate and exchange their credits gained from contributing food waste for free shared rides and free use of public facilities. In the whole system, the credit system plays a central role of incentivizing residents to make sufficient and sustained food waste contributions, so that they can continuously enjoy the communal benefits (i.e. higher mobility and more opportunities for social interactions).

While conceiving the sub-systems is crucial for developing the whole sharing system, outlining their inter-relationships helps to more firmly anchor the architectural and urban design interventions. As shown in Fig. 6.2, in-depth studies on the spatial distribution of food waste and anticipated demand of shared rides were conducted first. These distributive patterns then provided key references for planning the route of the self-driving cars, in order to achieve efficient food waste collection as well as an expansive catchment area. Following this, the proposed route became one of the main factors for locating the centralized digester and identifying appropriate pocket public spaces, where stations of personal mobility devices and other public facilities could be placed. At the same time, the planning of all these spatial components was coordinated with careful reorganization of traffic in the larger urban area.

Continuing with the systems approach, the larger environment in which the whole sharing system is nested was then carefully examined to further refine the design scheme. Again, taking the planning of self-driving cars for instance, the route mentioned above was in fact not a key design consideration at the very beginning, because the self-driving cars were originally proposed to operate freely across the entire neighborhood and fully on an on-demand basis. However, this idea turned out to be not as feasible as proposed when the larger environment was taken into account. The system of self-driving cars would ineluctably become an integral component of a larger traffic system in Singapore. This means that it is likely to impose additional pressures on the existing traffic network, presumably causing congestions and higher energy consumptions. As such, the intended objectives of higher mobility and energy efficiency presumed by the self-driving cars may not be achievable as a result of these undesirable consequences. It was this anticipated threat that led to the planning of the route, so that potential conflicts between the self-driving cars and existing traffic system could be contained, if not fully addressed.

The non-physical aspect of the larger environment is equally, if not more, important to informing the design decisions. Using the same example, in theory the catchment area of the self-driving car service is primarily determined by the core technologies (e.g. the furthest travel distance on a single charge). And this was exactly the basis on which the current route was calibrated and planned. Following this logic, with technological advancement, we can hypothesize that the present route can be extended to cover a much larger area and more destinations by employing a larger fleet of self-driving cars with higher charging capacities. And this would lead to greater mobility for many more people. However, this hypothetical objective may not be achievable and plausible. This is because not only the aforementioned

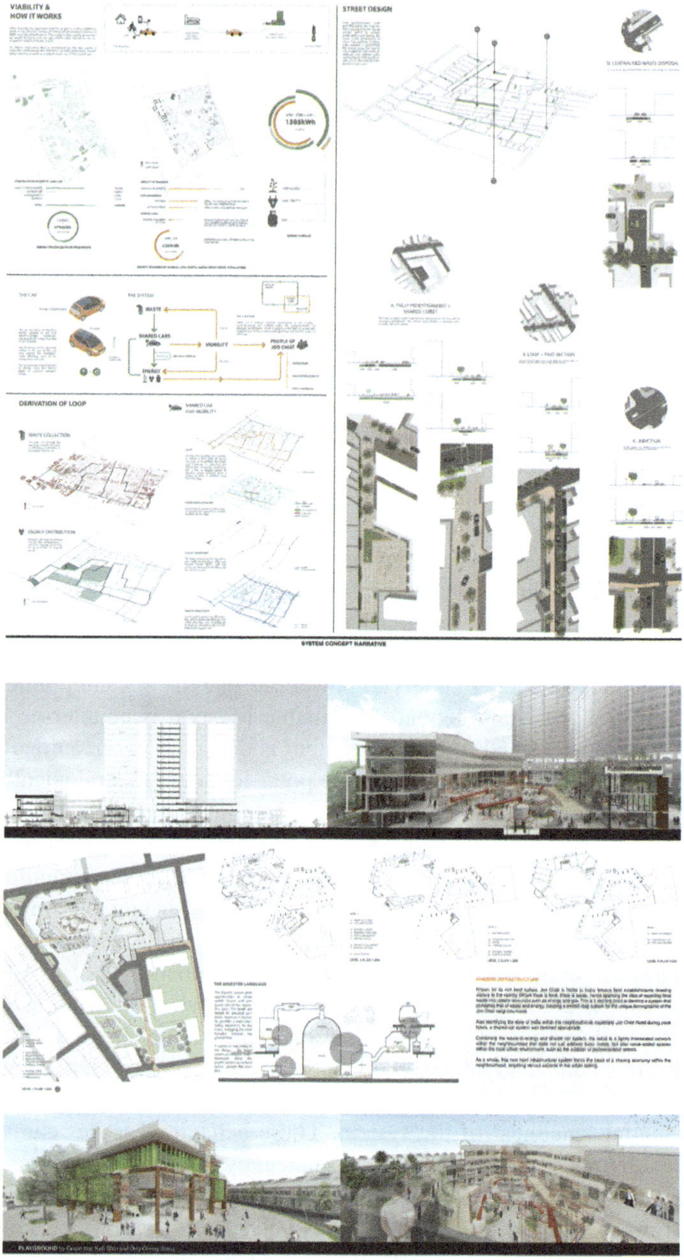

Fig. 6.2 Sharing infrastructure (*Source* Extracted from drawings of the NUS-Tsinghua Joint Studio: Sharing Cities)

increased burden on the existing traffic network is likely to be a severe hurdle, but the commensurate increase in energy demand due to longer travel distances and the growing number of self-driving cars also means that the system may exceed the tipping point, whereby the total amount of energy generated from all food wastes within the neighborhood becomes insufficient to power the entire fleet even if the waste-to-energy conversion ratio could be improved at the same time.

Although solutions such as expanding the geographical area of food waste collection could be considered, these solutions may nonetheless turn out to further undermine the very concept and the goal of the sharing system. After all, it is as counter-productive as it is unethical to increase the volume of waste in order to increase the shared mobility benefits brought about by these self-driving cars. One could conclude from this specific instance that while changing the parameters of the larger environment may offer new possibilities for design, the design itself could turn out to undermine the original intent if the constraints of this larger environment were carefully considered.

In addition, and staying in line with the systems approach, the design of the sharing system can be further tested and refined by looking into the possibilities of resource limitation in relation to people's behaviors. For this scheme, the following two scenarios are noteworthy. First, sustainable living in general aims to reduce consumption of resources, and when widely practiced, it is likely to result in an insufficient supply of food waste. In turn, this can lead to the malfunction of the proposed sharing system because of the lack of energy supply to power the shared cars and public facilities. In other words, a conscientious and sustainable lifestyle is counter-productive to this sharing system. Second, the credit system is designed to encourage residents' participation, but it may become a leverage point for some to strategically produce more food waste in order to earn more credits to enjoy more free shared rides and free use of public facilities. This means that a system founded on good intentions may ironically encourage perverse behaviors. For instance, a restaurant owner may run the business in a slightly less conscientious and less sustainable way in order to generate more food waste and hence accumulate more credits, which then allow this individual to offer more free rides to attract more customers.

In short, for the above two scenarios, the overall goal of the sharing system can ironically be undermined by individuals' conscientious under-production of food waste on the one hand, and on the other hand, the over-consumption of food and the abuse of free rides. From the perspective of the systems approach, both inclinations are the potential 'enemies' of the system. These behavioral issues can sometimes be addressed by refining the design, for instance, by introducing progressive credit reductions to the pre-empt opportunistic behaviors (e.g. where the amount of credit obtained from the second g of food waste is halved of that from the first kg). But design may after all be limited to address all these behavioral issues. Nevertheless, being aware of the 'enemy' of the system and its potential threats at least offers new pathways for reflecting on design, which is crucial for the design pedagogy.

6.6 Scheme Two: Sharing Open Education

The sharing system proposed in the second scheme is a shared open education platform. On this platform, both residents and visitors who have different expertise can offer related learning programs, which will constitute the future school curriculum on the one hand, and on the other hand, to facilitate continuous learning for the community. Through the production of this open education platform, this sharing system also aims to empower individuals and communities, and to create further opportunities for social interactions—the learning programs are anticipated to bring people of different backgrounds together and to enable them to share similar interests.

The sub-systems of this sharing system, and in particular their spatial components, were developed with a mindful consideration of and then a close relation to those of the first one. As shown in Fig. 6.3, the education platform comprises three hierarchies of physical components. The first hierarchy includes two major schools, which would be partially open to the public so that selected learning programs can be easily incorporated to enrich their curriculums. The centralized digester for electricity generation proposed in the previous scheme was also conceived as a learning base and integrated with one school. Altogether they constitute the key learning hubs within the neighborhood. Second, a number of underutilized public spaces across the neighborhood, including churches, mosques, temples, and gas stations, were identified and planned to be used as 'classrooms' for the learning programs on an on-demand basis. The last hierarchy comprises a selected group of the pocket public spaces and pedestrianized streets of the first scheme. Recycling facilities, small-scale digesters, and charging points for personally mobility devices would be installed in these spaces. This is to embed learning opportunities in the residents' everyday life.

Comparing the spatial layout of the two systems, it is not difficult to discover that most of the abovementioned components were deliberately located along the route of self-driving cars. In fact, the self-driving car route was also adjusted in view of the distribution of the schools and the underutilized public spaces. In this way, residents who subscribe to the learning programs can easily enjoy shared rides to travel from their home to the learning nodes, or between different learning nodes. And the learning facilities can also be more efficiently powered by the energy generated in the digesters and distributed by the self-driving cars in the form of batteries. In addition, the credit system proposed in the first system was also inherited to link these two sharing systems and to foster community participation. For instance, credit obtained from waste contribution can be used for subscribing to the learning programs. And credit would also be awarded to those offering and conducting programs based on the size and length of the class, and then used for enjoying free shared rides or for subscribing to other learning programs.

While the systems approach can help to build connections and to complement different sharing systems, it also sheds new light on the complex constraints and threats that one sharing system and its larger environment may impose on another. For instance, the technologies of waste-to-energy conversion and batteries not only determine the catchment area of the self-driving cars as described above, but they

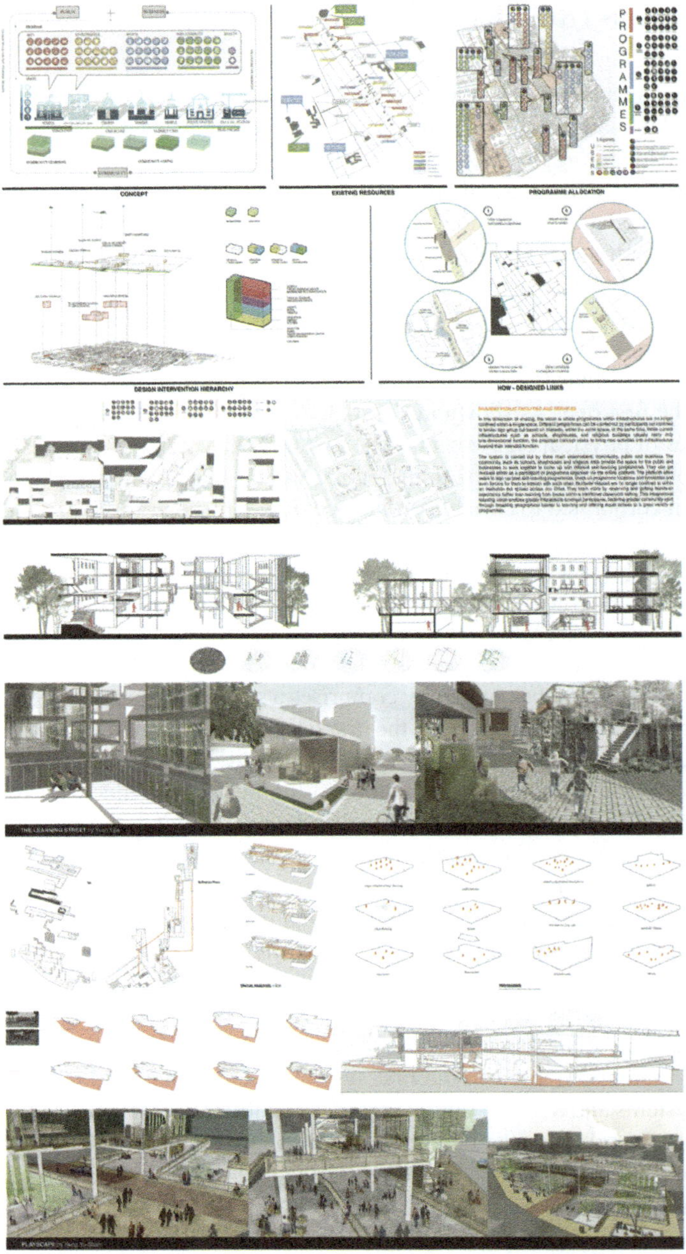

Fig. 6.3 Sharing open education (*Source* Extracted from drawings of the NUS-Tsinghua Joint Studio: Sharing Cities)

may also influence the quantity and the operation period of the learning facilities that are powered by the generated electricity. In addition, these learning facilities may also be underutilized for various cultural reasons or because of residents' biases, which in turn can undermine the goal of social inclusion. Conversely, if the learning programs are generally well received, then the increased demand of shared rides and more frequent use of the public facilities may impose a burden on the entire shared infrastructure system.

Applying the systems approach also offers insights on the 'enemies' of and between different systems. Again, consider the credit system as an example. Because of the above-described connections, earning more credits to participate in high-quality learning programs may become an additional incentive for some residents to produce more food waste. As a result, this is likely to further exacerbate over-consumption, which is one of the 'enemies' of the shared infrastructure system. At the same time, while good quality learning programs and reputable teachers are essential to building a successful open education platform, this may collectively result in the concentration of credits to a small group of people who are indeed professionally qualified to benefit from this sharing system. Nevertheless, for the whole community, this increasingly uneven distribution of credits may eventually discourage community participation—becoming the 'enemy' for both sharing systems.

A more detailed analysis of the two schemes would start by identifying other constraints and 'enemies' between the two systems. However, as stated before, analyzing the limitations and risks is not the ultimate goal of applying the systems approach to designing sharing systems. Instead, the aim is to continuously review the systems with new insights in order to refine the design proposals. By reviewing the two schemes, especially their interplays, new insights could be attained to improve the design of each individual ones.

We summarize in Table 6.2 the key considerations in the design of the sharing systems informed by the systems approach. Due to the scope of this chapter, this is only a partial depiction of the full systemic implications of these two schemes.

6.7 Discussion: Implication of the Systems Approach-Oriented Pedagogy for Design

The above two schemes demonstrate how the design of sharing systems and their practices can be guided and informed by the systems approach. What may be the rejoinders to the use of the systems approach in design? One may, for instance, argue that these schemes—such as waste-to-energy conversion facilities, emission free self-driving vehicles, or the regeneration of neglected public spaces—merely reflect the environmental focus observed in recent practices of architecture and urban design. And they may still be conceivable without applying the systems approach. How then, might a systems approach-oriented pedagogy be different, and more robust, than a

Table 6.2 The key considerations for the design of the two sharing systems and their practices informed by the systems approach

	Sharing Infrastructure	Sharing Open Education
Systems/sub-systems	• A fleet of electrical self-driving cars • A waste collection system as an integral part of self-driving cars • An energy generation and distribution system • A credit system	• The main waste digester of the first scheme and two schools that are partly open • Under-utilized public spaces and pocket public spaces across the neighborhood • A credit system shared with the first scheme
Objective/sub-objectives	• Energy efficiency • Enhanced mobility • Community participation	• Continuous learning • Community participation • Social inclusion
Constraints from the environment	• Waste-to-energy conversion technology • Battery technology for self-driving cars • Primarily organic waste	• Geographical proximity; • Cultural meaning of public places • Perceived biases of some public spaces • Language of courses
Threats to the environment	• More traffic congestions • Higher energy consumption • Reduction of efficiency • Reduction of convenience	• Conflict with existing education system • Conflict with existing programs of some public places
Enemies?	• Frugality and economical lifestyle • Over-consumption	• Reputable teachers and accredited programs • Certified private programs that compete with this program
Criteria for evaluating the system's performance	• Overall energy consumption • Social capital	• Inclusivity • Performance of understanding

Source Authors

conventional architectural design studio? Here, we suggest three different ways that a systems approach-oriented pedagogy is more robust.

First, the systems approach continuously pushes the boundaries of design investigation, allowing designers to broaden their scope of intervention, and to go beyond the architectural scale by positioning design as an integral part of a larger social system. While architectural and urban design remains a key focus, equal attention, if not more, has been placed on the interactions between socio-technical components. For instance, the credit system discussed in the case study can be seen as the central component in both schemes—a socio-technical component connecting different spatial interventions and an interface that links residents and their everyday lives to the entire system. This credit system might have been conceived through conventional architectural and urban design. But we suggest that it is unlikely to be in such a systemic manner; after all, deliberately focusing on any socio-technical component beyond the built environment is uncommon in architectural and urban design.

Second, the systems approach provides a multi-faceted prism, through which design investigation can be continuously and critically reviewed. In this process, more robust proposals can be developed by analyzing the different constraints and potentials, risks and opportunities, correlations, and consequences that are rendered visible by the systems approach. In contrast, it can be argued that conventional architectural and urban design pedagogies are often unjustifiably optimistic, and this can lead to the belief that design interventions always help to improve the overall situation. This same belief also leads to blind spots characterized by a severe lack of considerations of the boundaries, limitations, and in particular, the different 'enemies' intrinsic in any systemic design interventions. In particular, it is rare for a designer to either problematize or question his or her proposal in conventional architectural and urban design, which in turn may become the 'enemy' that can undermine the original intention. In contrast, the systems approach encourages critical questioning, and on this basis, empowers designers in making more informed design decisions, and in tandem, to reflect on the systemic impacts of these decisions.

Last, in addition to facilitating designers' capacities of dealing with complex design challenges, the systems approach also enables a different organization of the architectural design studio. For design studios attempting to explore complex topics that require collaboration and cooperation among students, the systems approach can be a very useful tool in structuring studio explorations. Since the end result of these studio explorations is usually a set of coordinated proposals, the systems approach can help to create a common and powerful platform for peer-to-peer learning. We found that throughout the entire semester, students frequently worked together in small groups, sharing design ideas and exchanging feedback. This groupwork practice is increasingly uncommon in architectural education because new technologies, such as networked laptops and portable 3D printing machines, offer students the freedom to develop their projects away from the studio. In contrast, we found that students' proposals were mutually informed, which we highlighted in the discussed schemes. And when compared to similar studios conducted previously without the application of the systems approach, these proposals were generally better developed with concrete research and more in-depth explorations.

Because the systems approach is a critical approach connecting different domains of knowledge, it can motivate students to explore other disciplines beyond the design studio. Through this exploration, students can come to discover new knowledge that will improve their capacities of addressing the aforementioned 'wicked' problems through design. This is especially crucial for the discipline of architecture and urban design, which does not have a clear disciplinary boundary and by nature is closely related to many other disciplines. It was also evidenced in our studio that in order to address the questions raised by the systems approach, many students became very proactive in consulting with us, and discussing their ideas with friends and classmates in other departments within the university. And for certain students, some even arranged regular meetings, tantamount to a regular but informal collaboration, for exploring new design solutions.

References

B. Balaram, The value of the sharing economy, in *Routledge Handbook of Sustainable Design*, ed. by R.B. Egenhoefer (Routledge, New York, NY, 2018), pp. 470–482

F. Bashier, Reflections on architectural design education: the return of rationalism in studio. Front. Archit. Res. **3**, 424–430 (2014)

S.M. Cahn, *Teaching Philosophy: a Guide* (Routledge, New York, NY, 2018)

C.W. Churchman, *The Systems Approach* (Delta Books, New York, NY, 1968)

D. Cuff, *Architecture: the Story of Practice* (MIT Press, Cambridge, MA, 1991)

N.D. de Graaf, D. Wiertz, *Societal Problems as Public Bads* (Routledge, New York, NY, 2019)

T. Irwin, The emerging Transition Design approach. Cuaderno **73**(19), 149–181 (2019)

A. Koch, K. Schwennsen, T.A. Dutton, D. Smith, *The Redesign of Studio Culture: a Report of the AIAS Studio Culture Task Force* (The American Institute of Architecture Students, New York, NY, 2006)

H.W.J. Rittel, M.M. Webber, Dilemmas in a general theory of planning. Policy Sci. **4**(2), 155–169 (1973)

S. Rohrbach, M. Steenson, Transition design: teaching and learning. Cuaderno **73**(19), 235–263 (2019)

D.A. Schön, *The Reflective Practitioner: how Professionals Think in Action* (Basic Books, New York, NY, 1983)

P. Scupelli, Teaching to transition design: a case study on design agility, design ethos, and dexign futures. Cuaderno **73**(19), 111–132 (2019)

G. Stevens, *The Favored Circle: the Social Foundations of Architectural Distinction* (MIT Press, Cambridge, MA, 1998)

Chapter 7
Conclusion

Sharing is an idea that we can all share despite our differences. We may disagree on what to share, how to share, and even why we ought to share. But all these disagreements presume that we share some common notion of sharing. And insofar as this notion is concerned, sharing is an ideal worth maintaining. In this book, we argue that it is possible to further amplify this ideal of sharing by design. Specifically, through the systems approach, we can design sharing systems that may narrow our various disagreements on sharing activities and practices, and in this process, accentuate the sharing ideal through material and organizational forms.

In this book, we begin by positing that the sharing economy, as it stands, is an inadequate platform and institution to develop the fullest social potential of sharing. Following this, we demonstrate the need for sharing by design. Subsequently, we describe the existing paradigms of sharing by design using 'design by capital' and 'design by commoning'. Our analysis suggests that each is limited in its own ways. In response, we offer an alternative paradigm of sharing by design using the systems approach. However, no account of the systems approach is ever complete without a discussion on the ethics of the system and how to transmit the knowledge embodied in this system to future generations. And so, we complete our account with a discussion on sharing ethics, which is followed by a reflection on teaching the design of sharing systems. At this point in the conclusion, we aim to briefly reflect on what we have done, and also what remains to be done. We will reflect along the following three main lines.

First, as system designers, we assume that by identifying and then specifying the basic components of the sharing system, and thereafter positioning them in interactive self-reinforcing feedback loops in relation to certain specified goals, we may start to approach the beginning of a sharing system. To describe this slightly differently, the system designer can assume the designed system to operate as he or she has foreseen. And while the systems approach is likely to concede to broader range of goals when compared to the paradigm of 'design by capital', as it is also likely to be a tad more structured and systemic relative to 'design by commoning' (see Chap. 3),

J. K. H. Chan and Y. Zhang, *Sharing by Design*, SpringerBriefs in Applied Sciences and Technology, https://doi.org/10.1007/978-3-030-43569-1_7

all these paradigms—including the systems approach—assume that a sharing system can be materialized by design. In making this assumption, it is important by systems thinking to also conjecture on aspects of sharing that do not fall into the purview of designers, or perhaps, cannot and should not be designed.

One aspect that comes immediately to mind is the virtue of benevolence. There are many plausible theories on why certain individuals are more predisposed to be benevolent while others are less so; but none of these theories can specifically explain where, or how, the virtue of benevolence emerges in sharing behaviors. We understand, especially from Widlok's (2004) work, that there were often norms for sharing, or communal obligations to share resources in traditional societies. But neither norms nor communal obligations can fully explain the presence of benevolence or compassion that drives many forms of sharing behaviors. On this, we suggest that while a well-designed sharing system is able to offer a conducive environment where moral behaviors are more rather than less likely, no amount of careful design deliberation and specification is able to create the virtue of benevolence in the sharing system. We cannot hope to design more virtuous individuals any more than we can try to instill the virtue of benevolence in sharing behaviors and relations. Benevolence, like many other virtues, is both an embodied moral quality as well as a freely chosen disposition of moral individuals. We may be predisposed to be benevolent; but very often, we choose to be benevolent. This kind of ethical choices cannot, and should not, come under the purview of design.

Beyond acknowledging this limit of design, it is also important to recognize another kind of limit. No design deliberation can fully foresee and anticipate all the emerging interactions, negotiations, and improvisations that in sum constitute the dynamic sharing practice. In such a practice, novel situations require swift adaptations and often drastic changes to ensure the continuation of sharing. This means that a robust sharing system ought to be designed with some open-endedness or incompleteness in mind—a possibility that Simon (1996: 162) anticipated as "designing without final goals". Permitting this open-endedness may mean enabling stakeholders of the sharing system to negotiate and to spontaneously reshape their own system to better fit their dynamic needs, where each change then creates a new situation that in turn offers a new starting point for fresh design activity (Simon 1996: 163). Clearly, designing for open-endedness or incompleteness is an epistemological and methodological challenge to both canonical systems design and conventional design methodologies, which presuppose some point of termination or closure in the design process. However, precluding this room for maximal openness and adaptive variation in design may mean restricting the sharing system in significant ways, which can result in its premature atrophy. The development of our systems framework in sharing by design does not yet offer a definitive solution to the challenge of an open-ended and incomplete design. We suggest that future research in the design of sharing systems may wish to focus on this challenging dimension of design.

Second, our rendition of sharing by design in this book has yet to offer a fuller commentary on the design of sharing power, which can be formulated simply as the political design of the sharing system. Although we have briefly commented on both participatory design and co-design in the preceding chapters, we have however assumed that power relations in sharing systems are largely symmetrical even when power relationships in the real world tend to be asymmetrical. Presumably, it is power that decides what is to be included or excluded in any exercise of systems design, and it is this same power that determines how to weigh or value the included components in this system. For Stavrides (2016), power is first and foremost the power to decide. In this way, the design of a sharing system has to presume power—yet power can neither be wholly captured nor discretely represented by any one component within the systems approach for sharing by design.

What is then a possible recourse? If we ascribe power to the stakeholders and the clients of the sharing system, we may over-commit power to these components and over-omit emergent sources of power from the interactions within a complex sharing system. But if we attribute power to the environment or the social forces of the sharing system, we may end up under-specifying power, which does not aid the task of design. For these reasons, we suggest that a possible recourse may reside in using the activity of systems design to foster the sharing practice itself. In other words, despite power differences, individuals come together to design the very sharing system that they would eventually inhabit. And through this process of designing the nascent sharing system together, vital capabilities and relations such as conflict negotiation and resolution, trust and deep friendships are formed. In turn, these newfound capabilities and relations then may come to counter-balance the worst excesses of power imbalance in the sharing system. Future research and practice of designing sharing systems may wish to experiment by embedding some activities of systems design within the formation of the sharing practice itself. To paraphrase Escobar (2017), the sharing community that practices the continuous design of itself may make it more likely for power to be shared—if only because this kind of systems design neither permits the centralization nor the consolidation of power.

Our third and final line of reflection focuses on methodology. In this book, we have presented the systems approach as a plausible design method for materializing a sharing system. To render this design method as clear as possible, we have emphasized on the most conceptual framing and implementation of the systems approach. However, to function well, the systems approach has to presume other design research methods, and it is important to ask the following key questions here: to what extent does the systems approach benefit from a mixed method design approach? And what are these other research methods that can complement the primary design method of the systems approach? In short, which design research methods are more robust bedfellows with the systems approach? If the overall aim of sharing by design is to provide a systematic primer for the design of a sharing system, then a more vivid picture to the underlying methods of design may be in order.

As discussed in the preceding chapters, the co-design method is often limited by fluid participation, and uncertain political dynamics. And this particular method, when well facilitated, may yield crucial data that can further inform the systems approach as the primary method for sharing by design. For example, Manzini (2015) suggests different methods of co-design spanning from mapping and visualizing information, to creating small-scale prototypes where users can come to experience the 'final' design and then communicating their feedback to other participants. In contrast to the more structural dynamics of the systems approach, the co-design approach offers an enlivening process of design—where stakeholders interact to produce ideas, constructive debates, and very often, also new data that can be used to improve the overall performance of the sharing system. And simultaneously in this co-design process, other research methods such as semi-structured interviews, questionnaires, and focus group discussions can be interleaved to draw out more structured or in-depth data that otherwise might be lost amid the fluid nature of a dynamic co-design process. The systems approach does offer an appropriate fit for the challenges of sharing by design. However, complementing the systems approach with other design research methods is likely to lead to an even more robust approach to sharing by design. Future research and design activity for sharing may wish to explore these methodological possibilities and potentials.

References

A. Escobar, *Designs for the Pluriverse: radical Interdependence, Autonomy, and the Makings of Worlds* (Duke University Press, Durham, NC, 2017)

E. Manzini, *Design, When Everybody Designs: an Introduction to Design for Social Innovation* (MIT Press, Cambridge, MA, 2015)

H.A. Simon, *The Sciences of the Artificial*, 3rd edn. (MIT Press, Cambridge, MA, 1996)

S. Stavrides, *Common Space: the City as Commons* (Zed Books, London, UK, 2016)

T. Widlok, Sharing by default? outline of an anthropology of virtue. Anthropol. Theory **4**(1), 53–70 (2004)

Index

J. K. H. Chan and Y. Zhang, *Sharing by Design*, SpringerBriefs in Applied Sciences and Technology, https://doi.org/10.1007/978-3-030-43569-1